ANDREW P. SAGE
Systems Management for Information Technology and Software Engineering

ALPHONSE CHAPANIS
Human Factors in Systems Engineering

YACOV. Y. HAIMES
Risk Modeling, Assessment, and Management, 2/e

DENNIS M. BUEDE
The Engineering Design of Systems: Models and Methods

ANDREW P. SAGE and JAMES E. ARMSTRONG, Jr.
Introduction to Systems Engineering

WILLIAM B. ROUSE
Essential Challenges of Strategic Management

YEFIM FASSER and DONALD BRETTNER
Management for Quality in High-Technology Enterprises

THOMAS B. SHERIDAN
Humans and Automation: System Design and Research Issues

ALEXANDER KOSSIAKOFF and WILLIAM N. SWEET
Systems Engineering Principles and Practice

HAROLD R. BOOHER
Handbook of Human Systems Integration

JEFFERY T. POLLOCK AND RALPH HODGSON
Adaptive Information: Improving Business Through Semantic Interoperability, Grid Computing, and Enterprise Integration

ALAN L. PORTER AND SCOTT. W. CUNNINGHAM
Tech Mining: Exploiting New Technologies for Competitive Advantage

REX BROWN
Rational Choice and Judgment: Decision Analysis for the Decider

WILLIAM B. ROUSE AND KENNETH R. BOFF (editors)
Organizational Simulation

HOWARD EISNER
Manging Complex Systems: Thinking Outside the Box

D1271682

MANAGING COMPLEX SYSTEMS

ABOUT THE AUTHOR

Since 1989, Howard Eisner has served as Distinguished Research Professor and Professor of Engineering Management and Systems Engineering at The George Washington University, Washington, DC. For the prior thirty years, he held various positions in industry, including president of two systems and software engineering companies (Intercon Systems Corporation and the Atlantic Research Services Corporation) and as a member of the board of three companies. He is a Life Fellow of the Institute of Electrical and Electronics Engineers (IEEE) and a member of several engineering honor societies. Dr. Eisner has written two books on systems engineering and related topics and a book on personal and corporate reengineering. He holds a B.E.E. from the City College of New York, and M.S. from Columbia University, and a Doctor of Science from The George Washington University.

MANAGING COMPLEX SYSTEMS
Thinking Outside the Box

HOWARD EISNER

WILEY-INTERSCIENCE

A JOHN WILEY & SONS, INC., PUBLICATION

Copyright © 2005 by John Wiley & Sons, Inc. All rights reserved

Published by John Wiley & Sons, Inc., Hoboken, New Jersey
Published simultaneously in Canada

No part of this publication may be reproduced, stored in a retrieval system, or transmitted in any form or
by any means, electronic, mechanical, photocopying, recording, scanning, or otherwise, except as
permitted under Section 107 or 108 of the 1976 United States Copyright Act, without either the prior
written permission of the Publisher, or authorization through payment of the appropriate per-copy fee to
the Copyright Clearance Center, Inc., 222 Rosewood Drive, Danvers, MA 01923, (978) 750-8400,
fax (978) 750-4470, or on the web at www.copyright.com. Requests to the Publisher for permission
should be addressed to the Permissions Department, John Wiley & Sons, Inc., 111 River Street, Hoboken,
NJ 07030, (201) 748-6011, fax (201) 748-6008, or online at http://www.wiley.com/go/permission.

Limit of Liability/Disclaimer of Warranty: While the publisher and author have used their best efforts in
preparing this book, they make no representations or warranties with respect to the accuracy or
completeness of the contents of this book and specifically disclaim any implied warranties of
merchantability or fitness for a particular purpose. No warranty may be created or extended by sales
representatives or written sales materials. The advice and strategies contained herein may not be suitable
for your situation. You should consult with a professional where appropriate. Neither the publisher nor
author shall be liable for any loss of profit or any other commercial damages, including but not limited to
special, incidental, consequential, or other damages.

For general information on our other products and services or for technical support, please contact our
Customer Care Department within the United States at (800) 762-2974, outside the United States at (317)
572-3993 or fax (317) 572-4002.

Wiley also publishes its books in a variety of electronic formats. Some content that appears in print may
not be available in electronic formats. For more information about Wiley products, visit our web site at
www.wiley.com.

Library of Congress Cataloging-in-Publication Data:

Eisner, Howard, 1935–
    Managing complex systems: thinking outside the box/by Howard Eisner.
      p.   cm.
    "A Wiley-Interscience publication."
    Includes bibliographical references and index.
    ISBN-13 978-0-471-69006-1 (cloth)
    ISBN-10 0-471-69006-6 (cloth)
    1. Systems engineering–Management.   2. Project management.   I. Title

    TA168.E387   2005
    658.4′04′0262 - - dc22   2005043921

Printed in the United States of America

10 9 8 7 6 5 4 3 2 1

Contents

Preface

This book is about two primary topics: systems and ways of thinking. We are part of and use several systems just about every day: from the highway system, to the telecommunication system, to the banking system, to the medical system, to many others. This central fact supports the notion that it is important to better understand how to build and manage such systems. Improvements can pay large dividends in all our lives and provide benefits measurable in economic as well as quality-of-life terms.

Such improvements are not, however, easy to make. Even if you are inside a company making a special product, it is often a long road from thinking about a product improvement to getting it out the door to your customers. The realities of how we conceive of and actually make new or improved products often conspire against us. So a large number of people in various companies and parts of government have put out the challenge for themselves and their associates—let's see if we can find a way to "think outside the box." This book is partly a response to that challenge.

The specific story of this book started some years ago when I was asked to teach a course in systems engineering at a relatively large systems integration and engineering company. About 30 employees were sponsored and made up a cohort that was being given this out-of-the-ordinary treatment. The course was a normal offering as the first course in systems engineering and part of my overall responsibility in the Engineering Management and Systems Engineering Department in the School of Engineering and Applied Science at The George Washington University. As lead professor in systems engineering, I decided to visit with the company manager who wanted this special program for his people. I asked him what he was looking for in the course and its delivery. He gave two answers. Since systems engineering was a core competency in the company's ability to do systems integration, he wanted to make sure that his people had a formal knowledge of systems engineering, along with some practice in execution. I more-or-less expected that answer, but then he proceeded to a second answer: "I want my people to be better able to think outside the box," he declared. I took that as a very serious answer, and it motivated me to dig more deeply into its various meanings.

My several-year research activity into what it might mean to think outside the box included a reexamination of my experiences over about a 45-year period. During the first 30 of these years I worked in industry, starting as a research engineer and finishing as the president of two high-tech companies. For the past 16 years I have been a professor in academia, where I served as an educator, researcher, and consultant to several companies and groups. Reviewing these experiences was enormously rewarding as well as lots of fun. It's not often that one gets a chance to review so many years of work with a focused objective in mind. A key question was: Could I identify specific cases where I concluded that I might be thinking outside the box, as well as to the contrary?

Along with the somewhat personal retrospective noted above, I also scoured the literature and found it expansive. Relevant material was everywhere and took me through a time line from da Vinci's era to the twenty-first century in terms of both technical and management thinkers. A few even took me back to our very earliest days as, for example, I contemplated what went wrong as some tried to build the Tower of Babel. I'm happy to report that we seem to have been able to do a lot better since that time.

As I was bringing together and organizing my personal experiences along with examples from our history and the literature, a list of some dozen ways to think outside the box emerged. All had at least some foundation in my own experience since I was able to connect to that perspective in a deeper way. In 1999 and 2000 I wrote two articles for the *Washington Business Journal* on the matter of thinking outside the box. Each article described briefly five ways to do this type of thinking. The reader response to these articles was about 10 times the response I typically got from my other articles on the "technology manager." This was additional encouragement; people certainly were very interested in this subject.

I followed that with a talk on this topic as the invited lecturer at an evening professional society meeting and dinner. Here again, a lot of interest was displayed, as I had the additional opportunity of a Q & A session as part of the evening activity. All of this convinced me that it was time to think seriously about how to write a book that included these experiences and my newly found set of expanded ways of thinking. The key to moving forward was to link the matter of thinking outside the box with my strong interest in systems and how we might be able to improve how we build and manage systems. I also wanted the book to be as useful as possible, sacrificing much of the academic perspective in favor of real-world practicality. Time will tell whether or not that goal has been achieved.

As to the particulars, this book is organized into 16 chapters. In the first four chapters we explore such topics as the nature of large-scale systems and complexity, thinking inside and outside the box with respect to specific systems issues, some basics of systems engineering and project management, system of systems engineering, typical systems problems faced by management, and examples of the inventive mind in both management as well as scientific domains.

Chapters 5 through 13 then focus on nine specific ways to think outside the box. Examples are used to a large extent to show how these thinking approaches have led to better solutions in the past. In this context, such solutions tend to contribute to improvements in how we build and manage systems of various types.

In Chapter 14, ways of thinking are expanded specifically to group situations. Such scenarios can be enhancers or inhibitors, depending on the group dynamics. Of particular interest is how to achieve the former and avoid the latter.

Chapter 15 widens the circle by presenting approaches that have been suggested by others to expand one's ways of thinking. These are not included as one or another of the main nine perspectives in Chapters 5 through 13 since the latter are derivative of more personal and direct experiences of mine. References are provided so that readers who wish to explore other approaches may do so.

Chapter 16 is the concluding chapter. It reiterates the main points and perspectives with respect to building and managing complex systems and thinking outside the box. It also includes a test for readers to obtain insights into whether or not they are ready to embrace thinking outside the box as well as some of the consequences of doing so.

Having looked at all of the above, a few words are still in order about who this book might be for. Several audiences are envisioned:

1. People with management responsibilities of various types with respect to building and managing large-scale complex systems
2. Nonmanagers trying to contribute to the building of systems, who are looking for new ways to approach the problems they encounter
3. People who are simply seeking to expand their patterns of thinking in any domain of their lives
4. University students who may be able, relatively early in their careers, to master new thinking perspectives that will help them be more successful in school and after they graduate

Relating to item 4 above, a standard 15-week university course may be constructed as follows:

Class Sessions	Book Chapters
1	1 and 2
2	3 and 4
3 through 11	5 through 13
12	14
13	15
14	16
15	Final exam

At a personal level, I wish to dedicate this book to four people. The first is my wife, June Linowitz, who continues to be supportive of my various writing endeavors and who, without much apparent effort, exhibits prodigious skills at thinking, both inside and outside the box. The other three people are my daughter, Susan Rachel Lee, and my sons, Seth Eric and Oren David. All are in their 40s and display considerable wisdom in conducting their lives. And they are passing this on to their children, my grandchildren, who are Gabriel, Jacob, Rebecca, Benjamin, and Zachary. The wheel continues to turn. Thank goodness it is (mostly) round rather than multipolygonic. There are still some bumps in the road.

HOWARD EISNER

Bethesda, Maryland

Chapter 1

Systems and Thinking

The main focus of this book is to define and explore ways of thinking that have been shown to have a positive impact on building and managing complex systems. This chapter is largely introductory, defining some basic aspects of systems as well as issues that relate to the construction and management of these systems. Some alternative approaches to thinking about these issues are also explored.

Complex systems will be interpreted very broadly and will include both physical (mostly hardware and software) groupings of equipment to serve a purpose and sets of procedures that are carried out by people or machines, or both. An example of the former is our national air traffic control (ATC) system that is used to control our aircraft as they fly across the country as well as in and out of terminal areas. For the latter, one might think of an airline reservation system, which is largely procedural but is executed on a large scale by both people and machines. Complex systems tend to be relatively large, with lots of internal and external interfaces. Part of our job in thinking about such systems is to try to reduce the complexity by focusing on fundamentals and simplification. But that is getting a bit ahead of ourselves.

Very large and complex systems include, as examples, the following:

- Our national aviation system
- Our national telephone system
- Our stock market system
- Our legislative system, the means by which we enact our laws

Managing Complex Systems: Thinking Outside the Box, By Howard Eisner
Copyright © 2005 John Wiley & Sons, Inc.

- Our electricity delivery system
- Our interstate highway system

Smaller, but still seriously complex systems could be represented by:

- Our traffic control systems, which control the flow of airplanes and automobiles
- The aircraft and automobiles themselves
- Information systems of various kinds, such as Microsoft Office or a company's inventory control system

It is important to note that some systems are related directly to a special or novel idea and that such systems would probably not exist without that idea. For example, when you send a FedEx package from the east to the west coast, you are using a large and complex system (mostly transparent to you) of airplanes, trucks, control systems, tracking systems, delivery personnel, and others. The fact that all packages go to Memphis, Tennessee, overnight and fan out from there is a fact of the design of that system. That design is based on the singular idea that this approach is efficient and cost-effective. Without that idea, another system configuration might have been selected. In other words, for many systems, it is the breakthrough idea that is behind the system, and as such, places us in search of these types of ideas as we approach the design issue. In this book, these types of ideas represent *thinking outside the box*.

1.1 BUILDING AND MANAGING COMPLEX SYSTEMS

We have a variety of tools that we use to perform the tasks of building and managing complex systems. In the building part, a set of activities known as *systems engineering* is dominant, giving us guidance on how to construct a cost-effective system [1.1–1.4]. A crucial part of that process is the architecting of the system, which is a top-level structure for the system [1.4, 1.5]. For example, the personal computer today generally has an *open architecture*, such that there is considerable interoperability between components. The same is true, in the main, for other consumer electronics, such as TV sets, VCRs, DVD players, and stereo (audio) systems. Unfortunately, there are lots of system architectures that do not adequately consider interoperability matters.

For the managing function with respect to complex systems, all of our institutions study and implement the best approaches they can think of in terms of management practices. There is an extensive literature that is available to help, from Jack Welch in regard to how he managed GE [1.6], to Harold Geneen and how he built ITT [1.7]. And there is no dearth of "gurus" such as Peter Drucker [1.8], Tom Peters [1.9], and W. Edwards Deming [1.10], who try to point us in the right direction with respect to the fine art of management. In relation to the field of project management, even I have joined the fray [1.4, Chaps. 3–6].

In both building and managing these systems, we are constantly in search of the new and better idea that will allow us to make critical improvements in their design and operation. The new and seminal idea is the "Holy Grail" that takes us to the next level, so that we may be in a position to do better than our competitors. Our overall industrial system works that way, and is supported in part by an elaborate system that is an example of how we value the new and better approach: our patent system. Without that system for recording and regulating better ideas, it would probably be a free-for-all in industry and a sure road to disaster. In other words, we continue to seek, keep track of, and value the better idea, referred to in this book as what results from thinking outside the box.

1.2   SOME RESULTS OF THINKING OUTSIDE THE BOX

In very specific terms, it is clear that thinking outside the box often leads to a new system that is better than previous systems. This can be illustrated by the following short list of inventions and better approaches that all will recognize:

- Light bulb
- Airplane
- Transistor
- Electronic chip
- Copying machine (i.e., xerography)
- Digital computer, from personal computer to supercomputer
- Atomic bomb
- Internal combustion engine
- Artificial heart
- Nuclear power plant

Although it is the new and seminal idea that sets the stage for major advances in building and managing systems, such an idea is often insufficient. Usually, there are many hurdles to jump over and potholes to avoid between the thought and its implementation. Bringing an idea from its conception to its fruition in the real world is often a long journey that requires lots of determination and stick-to-itiveness. A relevant observation in this regard is that genius is a matter of 1% inspiration and 99% perspiration (attributed to Thomas Edison). Although modest gains, one day at a time, can be made just by showing up, breakthrough gains don't come that easily. So even if you have a demonstrably wonderful new idea, you must keep in mind that you're still only a short way down the road toward the goal of converting the idea into a successful product or service that is implemented in the real world.

TABLE 1.1   Examples of Thinking Inside and Outside the Box

Issue	Thinking Inside the Box	Thinking Outside the Box
1. Integration of stovepipes	100% of all systems must be integrated	Integrate what it is cost-effective to integrate
2. Best of best of breed	Optimizing subsystem choices will optimize the overall system	May not work; there is no guarantee
3. Measurements	Measure as much as you can think of	Measure a minimum set that works and "tells the story"
4. Getting back on schedule	Add more people on the project	Adding more people is likely to make situation worse
5. Requirements change and volatility	Requirements are to be taken as fixed and inviolate	Requirements can, at times, be variables
6. Reserves on a project	All levels of management need to have dollar reserves	Project manager needs enough money to get the job done
7. Customer negotiation	Promise whatever the customer appears to want	Try to underpromise and overdeliver
8. Dealing with customers	The customer is always right	The customer can often be wrong
9. Overall approach	Do it right the first time (DIRFT) [1.11]	Provide continuous improvement and iteration
10. Employee trust	Employees cannot be trusted to know how the company is really doing	Have the obligation to tell the truth and focus on company well-being
11. Work task strategy	Never do work unless you can profit from it	Invest in key areas for the future health of the company
12. Processes and products	Get the process right and the products will always be right	The right process still doesn't guarantee the right product

1.3   THINKING IN RELATION TO SPECIFIC ISSUES

Table 1.1 lists a dozen issues together with ways of thinking about these issues that might be considered both inside and outside the box. These are explored next, one issue at a time. They are meant simply to illustrate, in a concrete way, how thinking inside and outside the box may differ from one another.

1.3.1   Integration of Stovepipe Systems

There are a large number of "stovepipe" systems in operation today, each of which carries out a discrete function. Examples include a human resources personnel tracking system, an inventory control system, and a finance and accounting system. When a new manager comes upon such a scene, he or she often leads the charge toward the integration of the stovepipes, believing that an integrated system is certainly going to be more cost-effective. The knee-jerk goal is often something like the following: We need to integrate the stovepipes to the maximum extent, approaching 100% integration. One might say that this knee-jerk reaction to the existence of a set of stovepipes is the rule today rather than the exception.

A deeper look, however, suggests that the integration of stovepipes may be a good idea, but it also may be a bad idea. Further, the so-called goal of 100% integration may be a *very* bad idea. An out-of-the-box approach might well be to integrate the stovepipes to whatever extent is appropriate and cost-effective, based on the specific circumstances attendant on the stovepipes in question. This author has seen an advanced Navy information system start down the road toward the integration of stovepipes and then be given up after three years, at which time they proceeded to "disintegrate" the stovepipes. The problem, as defined, was just too difficult and expensive.

More will be said about this particular out-of-the-box perspective later in this book. For the time being, can you think of situations for which the integration of stovepipes might well be a less than wonderful idea? If so, make a note of your thoughts and hold on to them until you read the next section.

1.3.2   Best of "Best of Breed"

Inside-the-box thinking assumes that a system composed of the integration of a set of best-of-breed systems will necessarily be the best system that can be constructed. After all, what can be better than the "sum" of "bests"? A simple example should demonstrate that such a system can be a terrible choice in the sense that another approach can easily be seen to be superior.

Let us assume that a government agency has done several in-depth studies to determine the best-of-breed systems that carry out certain functions. The functions and the best of breed answers are as follows:

Functions	Best of Breed Systems
Word processing	Wordperfect
Spreadsheet	Lotus 1-2-3
Presentation manager	Powerpoint
Database management system	Oracle

It should be noted that each of these best-of-breed systems is produced by a different company:

- Wordperfect is made by Corel.
- Lotus 1-2-3 is made by Lotus.
- Powerpoint is made by Microsoft.
- Oracle is made by Oracle.

Anyone familiar with software products recognizes the difficulty of trying to integrate software from different companies, even if the source code were accessible and modifiable. The costs of doing so would be prohibitively high, especially in light of the fact that another, more than feasible alternative is readily available. Those familiar with Microsoft's Office system (or Lotus's Smartsuite) can see immediately that an integrated system with the four functions above is available as a commercial off-the-shelf (COTS) product. Thus, abandoning the integration of disparate systems in favor of Microsoft Office is a clear and highly cost-effective alternative solution. This simple example demonstrates that leaping to the conclusion that a system composed of a set of best-of-breed systems will be the best choice may not be the right answer, even though one's intuition might point in that direction.

1.3.3   Measurements

This example focuses on measurements programs that are related to the building or managing of a product or service. Measurements are also sometimes called *metrics*. The inside-the-box approach is to measure everything that you can think of. That way you are not likely to miss anything important, one can reason. As an example, the Department of Defense, in relation to establishing *management indicators* for software, came up with the following measurement areas [1.12]:

1. Requirements volatility
2. Software size
3. Software staffing
4. Software complexity
5. Software progress
6. Problem change report status
7. Build/release content
8. Computer hardware resource utilization
9. Milestone performance
10. Scap/rework
11. Effect of reuse

Looking at this list, it is evident (1) that lots of measurements are being made, and (2) that it will take a lot of effort to make such measurements. The size and efficacy of a measurements program are, of course, related to the size of the system being developed, but still, the list of measurements above suggests a considerable effort at a very sizable cost. Outside-the-box thinking tries to focus on the *minimum* measurement profile that is sufficient to "tell the story" and be implementable within the constraints of a real-world situation. That includes the practicality of both making the measurements and making changes, given that the measurements suggest that there are problem areas.

This "minimalist" approach is supported, if you will, by some of the thinking in the ubiquitous and far-reaching Department of Defense. In the so-called 5000 series [1.13] dealing with the acquisition of systems, the directive and instruction of the reference are both scaled down, pointing to performance and capabilities-based acquisitions that are tailored to the situation at hand.

I tend to agree with this idea for most situations, reckoning that there should be an appropriate match between a measurement program and the specific needs of that program rather than going automatically to the default solution that says: "Measure everything that you can think of." A good measurements program should be designed rather than be the consequence of constructing a mindless, long laundry list.

1.3.4 Getting Back on Schedule

Many large-scale software development projects seem to go off the track after awhile, as manifested by falling behind schedule. This means that the actual achievement of various milestones is later than that called for by the original schedule. One inside-the-box reaction is to add personnel to the project in an attempt to get back on schedule. This reaction flies in the face of *Brooks's law* [1.14], which states that the addition of personnel will probably make the project schedule problem even worse rather than better. Part of the reason for this apparent anomaly is that productive people currently working on the project will have to stop what they're doing in order the bring the new people "up to speed." This diversion of their time and attention makes the schedule problem worse, among other negative effects.

Is there a potentially better solution to this type of problem? A few other approaches suggest themselves, including (1) changing the technical approach, which may not be sound; (2) recognizing that the present personnel level may not be sufficient in terms of the technical challenge and its degree of difficulty; and (3) the original schedule may not be realistic.

1.3.5 Requirements Change and Volatility

Inside-the-box thinking assumes that the up-front set of defined requirements is basically inviolate and "set in concrete." This point of view often follows from the need to be definitive in a contract document between customer and contractor. However, since requirements are usually defined very early in a program, it makes sense that they will probably have defects, often significant ones. Out-of-the-box thinking recognizes the fallibility of the people who develop this early statement of requirements and provides the means by which requirements can be changed in an organized and disciplined manner. This is not the same as requirements creep. Sensible changes to requirements should be made when both parties agree that such changes will improve the system development process or the product.

Support for the above can be found in both the "spiral" approach to building systems [1.15] and in aspects of the acquisition process in the government. With respect to the latter, we find the following phrase: "consistent and continuous definition of requirements." As systems are built, they are acquired in increments.

This incremental approach suggests further that requirements may change as part of the process, for purposes of refinement and improvement.

1.3.6 Reserves on a Project

Conventional wisdom might suggest, in the execution of a project, that each layer of management set aside a reserve to try to assure project success. That's an interesting mainstream idea but it can easily lead to counterproductive results. Suppose that a project manager (PM) reports to a program manager who reports to a division director who reports to a vice president. If the latter three executives each set aside a 10% dollar reserve, we can see that the PM is left with only about 73% of the original budget to work with. As a conscientious PM, he or she will then attempt to complete the work with about three-fourths of the original estimate of funding. This, in turn, can place enormous stress on the PM as well as on *every* member of the team over the entire duration of the project. This can easily lead to several negative consequences.

First, let us assume that by lots of extra hours and good PM management, the project is completed without use of the reserves. As the people on the project discover that all those extra hours were likely not to have been essential, they will conclude that they were taken advantage of, unless they get a bonus for the work. If not, a lasting negative feeling will persist.

Now assume that the project is not completed: that is, the PM runs out of money prior to completion. The reserves *may or may not* have been employed in time to preclude schedule and technical progress slippages. If not, a failure scenario has taken place, even though that was not the intention. The bottom line? Don't give the PM a very nearly impossible job to do and try to be a hero by making new funds available later. It's a strategy that has a good chance of backfiring. Instead, give the PM all the money necessary to get the job done.

1.3.7 Customer Negotiation

There are literally hundreds of books that can be consulted about how to negotiate with your customer, who is looking to you, and possibly others, to build and manage a system for them. Conventional wisdom appears to tilt in the direction of promising essentially whatever the customer wants. It is easy to see that this strategy can often lead both parties astray.

An alternative out-of-the-box approach was suggested to me by my oldest son. He had been quite successful as a software development manager at a large and significant company. One day I asked him to cite the most important reason for his success. His answer was: "Try to underpromise and overdeliver." This meant relatively hard-nosed up-front negotiation, but it paid handsome dividends to both parties down the road. At project completion, everyone tended to be pleased. Another way of expressing this approach is simply to *manage expectations*. Many managers work themselves and their people to a frazzle and still wind up with an unhappy customer. The reason: They haven't paid attention to managing expectations.

1.3.8 Dealing with Customers

Related to the above is the inside-the-box notion that the customer is always right. Since the customer consists of one or more fallible human beings, it is clear upon further consideration that the customer can be wrong at times, especially in relation to a nonmass marketing and sales situation. In a more conventional one-on-one context, your customer, for example, may:

- Desire that you meet an impossible schedule
- Not be willing to pay for the true product or project costs
- Ask for work that is outside the scope of the contract
- Not provide adequate customer-promised equipment
- Try to upgrade product specifications and requirements midway through a project, with no schedule or cost consideration

Any one of the above (and potentially lots of others) may lead you down the road to failure and should therefore be resisted. The perspective that the customer is always right may be a good metaphor for how generally to treat a customer, but it should not be accepted literally in most situations. Knowing when the customer is right, and when not, can easily translate into the difference between your success and your failure. One of the precepts of building complex systems is to walk away from a customer who insists on driving the relationship to a "win–lose" conclusion, with them on the winning side.

1.3.9 Overall Approach

There is a notion that comes from the field of total quality management (TQM) that might also be applied to one's overall approach to building and managing systems. This notion, coined by Philip Crosby [1.11], can be stated succinctly as "Do it right the first time" (DIRFT). Although that is not an undesirable thing to do, I would still put it in the category of being inside the box. What, then, is outside the box? It certainly could not be "Do it wrong the first time"!

In today's world of complex systems, it is necessary to acknowledge that the chances of doing everything correctly the first time is vanishingly small. So the question becomes: What approach should be taken in the light of this fact? The answer that lies outside the box in this connection is to focus on continuous improvement and iteration (CII). This perspective is from both the fields of TQM and systems engineering. The CII approach recognizes that we learn more about the system as we proceed through its development cycle, and what we know needs to be used to update earlier versions of the system. This relates to the discussion above about changing requirements, the "spiral model," and the value of iteration in order to converge on a system architecture and detailed design. This is not an invitation to make mistakes, but rather, to understand how to proceed when you are faced with unknowns and hurdles that need to be dealt with in a systematic and progressive manner. This is also reinforced by a key idea of Senge's [1.16]: that organizations need to become learning entities, which will better allow them to tackle matters related to continuous improvement and iteration.

1.3.10   Employee Trust

In the context of a project whose purpose is to build or manage a system, the project personnel are often not trusted to know fully the status of the project, especially if there is trouble. This reflects the idea that bad things might happen if these personnel are too aware of problems. Indeed, it also reflects the perspective that members of your team are not necessarily to be trusted to know the truth. This attitude is distinctly inside the box and should be replaced by an open and truthful approach to members of the project team. It is also perfectly all right to ask project personnel to keep any and all information about the project within the project team and *only* the project team. It turns out that people will respond very well to being trusted and not well to not being trusted. Isn't that true of you?

1.3.11   Work Tasks Strategy

With a project orientation, as discussed above, there is often the question of how to approach work tasks that are in the gray area, defined here as the set of tasks that might or *might not* be essential to successful completion of the project. How much effort, if any, should be applied to tasks within this zone of uncertainty? The inside-the-box answer tends to be guided by the admonition never to do work unless you can make a profit from it. This can get into the matter of the type of contract under which the work is being performed. Independent of that, however, is the out-of-the-box perspective that it is necessary to make certain investments that tend to be for the overall health and greater good of the enterprise, even though it might be detrimental to the budget of a particular project. The PM should be aware of the "greater good" scenario and bring such situations to the attention of his or her boss as soon as possible.

   In a context larger than that of a project, the same type of question can be addressed at the enterprise level. Companies need to invest in their futures, and the failure to do so might result in no future at all, in favor of maximizing profits for today (or the next quarter). Enterprises that are prosperous today do not necessarily have a license to that prosperity forever. A brief look at the company Wang Laboratories demonstrates that point: a multibillion-dollar and highly successful company in the 1980s that went into bankruptcy into the 1990s. They may have been investing in their future, but the way they did it is certainly open to question. The same may be said for quite a few "systems" organizations. The bottom line here is that all enterprises need to be open to investing in key areas, at all levels in the company, from the corporate level to the project level.

1.3.12 Processes and Products

There appears to be an inside-the-box perspective that if the processes are correct, it follows that the product will be also, and without question, will be correct. Processes are clearly important, but the matter of building and managing systems is not totally a function of these processes. The belief that "process is king" may lead to an inappropriate allocation of effort and funding that works against success instead of for it.

An outside-the-box view is that the correct process is a necessary, but not a sufficient, condition for success. In other words, there are other key ingredients to success, the most important of which is subject matter expertise. Teams with excellent process behavior may still lack the proper subject matter knowledge to both build and manage complex systems. Clearly, one should not expect a highly capable process-oriented team of communication engineers to build an excellent nuclear power plant, and conversely. Every enterprise needs to find the right process but must also make sure that each team has the right mix of subject-specific knowledge as part of the team expertise.

1.4 CONCLUSION

The issues discussed above have been examined from the perspective of what appears to be inside the box and what outside. Several of these issues will resurface in later chapters. Readers may or may not agree with all that has been said, and to a certain extent, that is expected. After all, outside-the-box views are, by definition, in the minority. If you are now itching to take issue with one or more of the dozen examples above, I urge you to be patient, as more will be said about that in later chapters. For the time being, let us look more deeply into how we approach the matter of building and managing complex systems and their ingredients. These are the main topics of Chapter 2.

REFERENCES

1.1 Sage, A. P. (1992). *Systems Engineering*. New York: Wiley.

1.2 Sage, A. P., and J. E. Armstrong, Jr. (2000). *Introduction to Systems Engineering*. New York: Wiley.

1.3 Blanchard, B., and W. Fabrycky (1998). *Systems Analysis and Engineering*, 3rd ed. Upper Saddle River, NJ: Prentice Hall.

1.4 Eisner, H. (2002). *Essentials of Project and Systems Engineering Management*, 2nd ed. New York: Wiley.

1.5 Rechtin, E. (1991). *Systems Architecting*. Upper Saddle River, NJ: Prentice Hall.

1.6 Welch, J. F., Jr. (2001). *Jack—Straight from the Gut*. New York: Warner Books.

1.7 Geneen, H. (1984). *Managing*. New York: Avon Books.

1.8 Drucker, P. F. (1986). *The Frontiers of Management*. New York: Harper & Row.

1.9   Peters, T. J., and R. H. Waterman, Jr. (1982). *In Search of Excellence*. New York: Harper & Row.

1.10 Deming, W. E. (1982). *Quality, Productivity and Competitive Position*. Cambridge, MA: MIT Press.

1.11 Crosby, P. (1984). *Quality Without Tears*. New York: New American Library, Penguin Books.

1.12 U.S. Department of Defense (1994). *Software Development and Documentation*, Military Standard 498. Washington, DC: DOD.

1.13 U.S. Department of Defense (2003). *Operation of the Defense Acquisition System*, Directive 5000.1 and Instruction 5000.2. Washington, DC: DOD, May 12.

1.14 Brooks, F. P., Jr. (1975). *The Mythical Man-Month*. Reading, MA: Addison-Wesley.

1.15 U.S. Department of Defense (2003). *Operation of the Defense Acquisition System*, Instruction 5000.2. Washington, DC: DOD, May 12.

1.16 Senge, P. M. (1990). *The Fifth Discipline*. New York: Doubleday Currency.

Chapter 2

Building and Managing Systems

It is assumed that many of the readers of this book are involved in some aspect of building or managing complex systems. Another assumption is that the reader has a strong interest in *improving* the manner in which such systems are built and managed. Doing better today than we did yesterday is a powerful motivation for thinking outside the box, and helping to achieve these goals is a major purpose of this book. However, before we move forward along these lines, it would seem necessary to articulate some of the fundamentals of building and managing complex systems, which is the immediate focus of this chapter. As a way of exploring these fundamentals at a top level, we define three areas as critically important in building and managing complex systems:

1. Systems engineering
2. Project management
3. General management

The third area is an extremely broad topic and aspects of it are discussed in Chapter 4 in relation to management gurus who had something special to say; the former two areas are explored briefly in this chapter. Also, in later chapters we continue to examine many specifics as to how thinking outside the box fits into these three domains.

Managing Complex Systems: Thinking Outside the Box, By Howard Eisner
Copyright © 2005 John Wiley & Sons, Inc.

2.1 SOME BASICS OF SYSTEMS ENGINEERING

Systems engineering may be thought of as a primary means by which complex systems are designed and built. A very brief definition is [2.1]:

> Systems engineering is an interdisciplinary approach and means to enable the realization of successful systems.

Sage and Rouse have provided structural, purposeful, and functional definitions of systems engineering [2.2], the latter of which is:

> Systems engineering is an appropriate combination of the methods and tools of systems engineering, made possible through use of a suitable methodological process and systems management procedures, in a useful process-oriented setting that is appropriate for the resolution of real-world problems, often of large scale and scope.

Yet another definition, from my own book [2.3], is:

> Systems engineering is an iterative process of top-down synthesis of a real-world system that satisfies, in a nearly optimal manner, the full range of requirements for the system.

One way to visualize the structure of systems engineering is depicted in Figure 2.1, which shows two levels of design: a set of system construction activities and a foundation consisting of 30 elements (listed later in this section). The two design levels are architecting and subsystem design. The former is a top-level design that selects, from among a set of alternatives, a basic approach to the system structure. Subsystem design continues in the design process, but at lower levels. This combination of two sequential design processes may be thought of as similar, in many ways, to what an A&E firm does: architecting and engineering.

My specific approach to the *architecting* process is described in Chapter 8. For now, we simply acknowledge its central position in the matter of systems engineering and note that if done correctly, it can serve to simplify the systems engineering process, thus coping with the vagaries of system complexity. In effect, a clear and consistent architecting process will be a major tool in cutting through the underbrush of system complexity. Very specific and key steps in the architecting method are:

1. Development and analysis of system requirements
2. System functional definition, decomposition, and allocation of requirements
3. Synthesis of alternatives
4. Analysis, evaluation, and presentation of alternatives
5. Selection of preferred alternative

Recognizing the importance of architecting, after many years of struggling with this issue, the Department of Defense defined an architectural framework in the

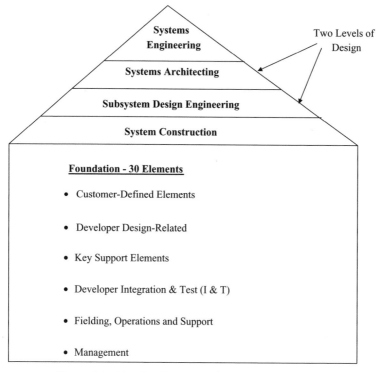

Figure 2.1   Top-level aspects of systems engineering.

command, control, communication, computer, intelligence, surveillance, and reconnaissance [C4ISR] domain [2.4].

Additional elements of systems engineering, referred to in Figure 2.1, include the following items [2.3]:

Customer-Defined

1. Needs, goals, and objectives
2. Mission definitions
3. Requirements
4. Functions to be accomplished

Developer Design-Related

5. System architecture
6. Analysis and evaluation of alternatives
7. Technical performance measurement
8. Life-cycle costing

9. Risk analysis and mitigation
10. Hardware, software, and human engineering

Key Support Elements

11. Concurrent engineering
12. Specification development
13. Interface control
14. Computer tool use
15. Technical data management and documentation
16. Integrated logistics support
17. Reliability, maintainability, and availability
18. Quality assurance
19. Configuration management
20. Specialty engineering
21. Preplanned product improvement

Developer Integration and Test

22. Integration
23. Verification and validation
24. Test and evaluation

Fielding, Operations, and Support

25. Training
26. Production and deployment
27. Operations and maintenance
28. Operations evaluation and reengineering
29. System disposal

Management

30. Management of all the elements listed above

The essential notion is that these procedures and steps will serve as a road map for the successful construction and fielding of a complex system.

2.2 SOME BASICS OF PROJECT MANAGEMENT

Many aspects of building and managing a complex system are executed within the context of a *project*. Hence, the field of *project management*, representing all the important activities that are performed as a team builds and manages a system, has become critically important. Whereas systems engineering tends to focus on *technical activities* (although not exclusively), project management emphasizes *management activities*. The combination of systems engineering and project management tends to define the centerpiece of an endeavor known as *systems integration*.

There are several top-level definitions of project management [2.5–2.7]. Perhaps the most prevalent is that this subject consists of (1) planning, (2) organizing, (3) directing, and (4) controlling. My definition [2.3] considers the fourth item above to be *monitoring*, on the grounds that an appropriate combination of directing and monitoring will provide whatever control is necessary. There is an extensive literature on the subject of project management, including a "body of knowledge" set of documents that can be accessed through the well-known Project Management Institute [2.8].

Most larger projects can be thought of as being managed by a *triumvirate*, with the chief systems engineer (CSE) and the project controller (PC) reporting to the project manager (PM). The CSE is responsible for technical matters, the PC is concerned with administrative matters (including project schedule and costs), and the PM accepts overall responsibility for all that goes on within the project [2.3].

The project can be an organic unit within an organization, and it carries out concrete pieces of work, often including the design and construction of a system. In a matrixed organization, the personnel for a project may contain at least some people for the project from a functional organization that might, for example, include groups of software engineers, hardware engineers, test engineers, and other engineering disciplines.

A key road map for a project is its *project plan*, consisting of at least the following elements [2.3]:

1. Needs, goals, objectives, and requirements
2. Task statements
3. Technical approach
4. Schedule
5. Organization and staffing
6. Budget
7. Risk analysis

There are several ways in which tasks may be described, including item 2 above, statements of work and work breakdown structures. The organization and staffing usually address the specific assignments of people to tasks, in the form of a task responsibility matrix. Once a project is under way and real costs are being expended in the various task areas, budget items expand into cost analysis.

2.3   COMPLEX SYSTEMS

Complex systems clearly constitute more of a technical and management challenge than do simple systems. In that sense, complex systems require stronger and more insightful thinking patterns to be successful. This provides a strong motivation for at least a brief examination here of some of the features of complex systems.

The consideration of complex systems has attracted a large number of competent investigators over many years. This has ranged from the work of a 1977 Nobel laureate, Ilya Prigogine [2.9], to relationships with the information-theoretic notions of entropy [2.10], to chaos theory [2.11, 2.12], to systems architecting [2.13], and to other significant areas, such as simulation [2.2]. The brief discussion here is focused more highly on aspects of systems that tend to be positively correlated with system complexity from an engineering perspective.

Systems tend to become more complex with each new version, despite our occasional interest in simplification. That is, we normally like to add increasing capability into our systems, and that makes them more complex. For example, our automobiles are considerably more complex today than were those coming off Henry Ford's assembly line at the beginning of the twentieth century. But they are also a lot more capable in just about every dimension. At the same time, even though we have electric can openers and pencil sharpeners, manual versions of each are still being sold, demonstrating that there is still a demand for a "simple" approach to life.

With the proliferation, in general, of more complex systems has come the need to find better solutions in both technical and management domains. This is a challenge that faces us today, and it is not likely to go away soon. A better understanding of what it is that leads to complexity, however, will help in that regard. The discussion below cites some factors that are associated with system complexity.

2.3.1   Size

System complexity tends to increase with system size. If we look at the air traffic control (ATC) system as an example, we note that it is a very complex system consisting of radar systems, communications systems, navigation systems, landing systems, and lots of other subordinate systems. This overall ATC system is certainly more complex than any of its subsystems. The same is true for many very large systems, such as other aspects of our national transportation and telecommunication systems.

2.3.2   Functionality

The complexity of systems also tends to increase as we demand that they carry out new functions. Not very long ago we had simple unifunctional software systems that could be described as (1) database management systems, (2) spreadsheets, (3) word processors, and (4) presentation managers. Now these four functions have been crunched together so that all are available as part of a larger system such as Microsoft Office and Lotus's Smartsuite. This clear increase in functionality has made the software systems more complex, and this basic idea holds for many other systems, including your alarm clock, which now plays AM and FM radio as well as compact disks.

2.3.3   Parallel vs. Serial Operation

System complexity also tends to increase as we demand more parallel operations rather than purely serial operations. Compared with serial operations, doing tasks in parallel implies more bandwidth and more separate components. Other words that are used in this respect are *multitasking* and *parallel processing*. This type of capability shows up in lots of systems: from simple personal computer operations to a complicated radar system that is able to keep targets in a track-while-scan mode while it is looking for more targets.

2.3.4   Number of Modes of Operation

Another correlative variable is the number of modes of operation that a system can support. When your desktop computer is not turned off correctly and then is turned on later, it usually goes into a "safe" mode, which has its own set of features, different from those of the normal mode of operation. Many wristwatches today have lots of modes of operation, allowing one to operate full and degraded modes as well as different functions that are part of the overall design (e.g., start and stop a timer, convert into a calculator). This aspect of system design can also be thought of as being connected directly to the functionality of the system. You are also likely to identify both with the "features" of a system, especially when your telephone company calls and tries to convince you of the merits of its caller-ID capability.

2.3.5   Duty Cycle (Dynamic vs. Static)

Duty cycle, which refers to a system's profile of use and the number of times that it switches from one state to another, is another aspect of a system that tends to be related to its complexity. High-duty-cycle systems are operating dynamically all the time, and low-duty-cycle systems can be resting (off) a lot of the time, and get turned on occasionally. An office copier is usually a high-duty-cycle machine, but the same copier in a home environment is likely to experience a lot less demand and fewer overall cycles per unit of time. In this case the machine is the same, but the demand makes the machine environment and profile of use more complex.

2.3.6   Real-Time Operations

A system that is required to respond in real time is generally more complex than one that does not have to do so. Most command and control systems from the military arena have a real-time requirement, especially when they are involved directly in warfare. If one is to shoot at an enemy missile that is coming at you on a ship, a clear sense of real-time urgency dominates the scenario. On the other hand, processing payroll overnight to produce checks for all employees can also be considered urgent, but if it doesn't run at midnight, one can try again at 4 in the morning.

2.3.7   Very High Performance

A system that has a very high performance demand can often be considered as complex. Here we are referring to systems that are pushing the state of the art and are at the edge of what we comfortably know how to do. For example, the air traffic control (ATC) system can be thought of as a very high availability system since we require that the system be in an "up" state a very high percentage of the time. We do not want to have aircraft subject to positive control from the ground to be faced with a lack of such control, with no backup system to call on in the event of failure. We often use online redundancy to achieve high availability, but this type of system is more complex than one without such a stringent requirement.

2.3.8   Number of Interfaces

A system with a large number of internal and external interfaces may be said to be more complex than one that does not have such a feature. This factor is likely to be highly correlated with size and functionality; none the less, it is something to contend with. We also know that many systems tend to fail at their interfaces, so we spend extra time and effort to make sure that they are designed and implemented correctly, and with low likelihoods of failure. Naturally, the greater the number of interfaces, the larger the size of the problem.

2.3.9   Different Types of Interfaces

The type of interface is also important with respect to complexity. Many systems have a simple physical (mechanical) and electrical interface, such as plugging together stereo components and hooking up a cable to a DVD player, VCR, or TV set. If we add thermal, environmental, data structure, and protocol interface requirements, the system becomes even more complex.

2.3.10   Degree of Integration

The degree of integration relates to the stovepipe issue, cited in Chapter 1. If there is a requirement to integrate many stovepipe systems, more complexity is created than in a system with a less stringent need for integration. Moving from zero integration to close to total integration increases schedule, cost, and complexity. Selecting the "best" level of integration is itself a complex matter that requires a deep analysis of the particulars of the situations (i.e., there is no simple answer that works for all types of systems). Down the road, it is hoped that a "degree of difficulty" index will be developed that will give us a better idea of when to stop trying to find a more highly integrated solution as well as when we already have a very cost-effective solution at hand.

2.3.11 Nonlinear Behavior

Many systems are highly nonlinear in their behavior, and such behavior may be said to be more complex than the well-known alternative of a (mostly) linear system. This is certainly true from a mathematical point of view, as our tools for analyzing nonlinear systems are less complete than those for linear systems. Further, our intuition about highly nonlinear systems often fails, due to their complex nature. Formal complexity theory highlights nonlinear behavior patterns that are part of the relatively new field of chaos theory.

2.3.12 Human–Machine Interaction

The human–machine interaction element of system complexity has been included to emphasize the role of the human being in building, managing, operating, and maintaining complex systems. A person acting in one or more of the latter roles can surely make a system better. But it is also true that a person can commit errors, and to combat these errors we often design and build a more complex system. For example, we believe that human error is one of the main reasons for automobile accidents. We also add features that will preserve life in the event of such errors (e.g., air bags). These features inherently make the system more complex. Greater complexity in the service of safety, however, is certainly a good idea.

Despite our attempts to the contrary, it is virtually an axiom that complex systems are more susceptible to failure than are simple ones. Noted earlier is our design to defeat such behavior through redundancy, generally a recommended solution. However, perhaps a more fundamental approach is to heed the old saw: K.I.S.S. (keep it simple, stupid) [2.14]. Is there a bottom line? A reasonable answer might be: Avoid complexity when it is not essential, but employ complexity in the service of satisfying true system requirements with a cost-effective solution. Are there practical ways that this can be done? We hope readers will recognize some of these ways in later chapters.

2.4 SYSTEM OF SYSTEMS ENGINEERING

A particular type of complex system is one that results from integrating a set of complex systems, thereby constructing a system of systems (SoS). The national air traffic control (ATC) system cited earlier as a complex system, consists of a variety of subordinate systems that perform several needed air traffic control functions (e.g., communications, navigation, identification). However, not only is the ATC system a system of systems, it is also a part of a larger SoS: the national air transportation system. The latter system adds additional subordinate systems, such as the set of airports, access and egress systems to and from airports, aircraft of many types, and others.

The matter of creating systems of systems has been closely linked to the problem of how to carry out complicated systems integration tasks. This has been referred to in this chapter and in Chapter 1 as the stovepipe issue (see Table 1.1). In short, when we attempt to integrate stovepipe systems, we also tend to create a system of systems. This can have an extremely broad range of degree of difficulty connected with it, and determining the degree of integration to target as a goal is arguably one of the more complicated systems engineering issues with which we must contend.

The literature regarding systems of systems and SoS engineering has been building, and it provides some insight as to how we might be able to deal with our tendencies to create these larger and more complex systems. Sage and Cuppan examined both the systems engineering and the management of systems as well as *federations* of systems [2.15]. Maier has suggested some principles for the architecting of SoS [2.16]. Charles Keating and several of his university colleagues have provided some top-level perspectives regarding the "concept, foundations, research directions, and practice implications for" SoS engineering [2.17]. In concert with others, I have set forth a way to explore the engineering and management aspects of SoS engineering [2.2], which are broken down into the following three topics: (1) integration engineering, (2) integration management, and (3) transition management. Further, each of the above consists of several task areas that must be dealt with in addressing the overall subject of SoS engineering (discussed in greater detail in Chapter 5). An additional expansion of these notions deals with possible ways to carry out this type of engineering so as to develop systems more quickly. These matters are considered within the overall topic known as the *rapid computer-aided system of systems engineering* [2.3]. It can certainly be expected that an increasing amount of attention will be paid to the matter of systems of systems as we move through the twenty-first century.

2.5 SUMMARY

Building and managing large-scale, complex systems is one of our continuing great challenges. As systems grow they become more complex and are often transformed into systems of systems. We rely on three primary disciplines to assist and guide us in dealing with such systems: (1) systems engineering, (2) project management, and (3) general management. However, as these fields become better refined and more effective, many of the problems that we face will continue to persist. A sampling of such problems from four perspectives—systems, people, software, and management—is explored in Chapter 3.

REFERENCES

2.1 International Council on Systems Engineering, www.incose.org.

2.2 Sage, A., and W. Rouse (eds.) (1999). *Handbook of Systems Engineering and Management*. New York: Wiley.

2.3 H. Eisner (2002). *Essentials of Project and Systems Engineering Management*, 2nd ed. Hoboken, NJ: Wiley.

2.4 See Command, Control, Communications, Computer, Intelligence, Surveillance and Reconnaissance (C4ISR) Web site: www.c3I.osd.mil.

2.5 Kerzner, H. (2003). *Project Management: A Systems Approach to Planning, Scheduling and Controlling*. Hoboken, NJ: Wiley.

2.6 Forsburg, K., H. Mooz, and H. Cotterman (2000). *Visualizing Project Management*. New York: Wiley.

2.7 Klastorin, T. (2004). *Project Management: Tools and Trade-offs*. Hoboken, NJ: Wiley.

2.8 Project Management Institute (2000). *A Guide to the Project Management Body of Knowledge*. Newton Square, PA: PMI; www.pmi.org.

2.9 Nicolis, G., and I. Prigogine (1989). *Exploring Complexity: An Introduction*. New York: W.H. Freeman.

2.10 Pincus, S. (1991). Approximate Entropy as a Measure of System Complexity, *Proceedings of the National Academy of Sciences*, Vol. 88, March, pp. 2297–2301.

2.11 Gleick, J. (1987). *Chaos: Making a New Science*. New York: Viking Penguin.

2.12 Waldrop, M. M. (1992). *Complexity: The Emerging Science at the Edge of Order and Chaos*. New York: Simon & Schuster.

2.13 Rechtin, E., and M. Maier (1997). *The Art of Systems Architecting*. Boca Raton, FL: CRC Press.

2.14 Rechtin, E. (1991). *Systems Architecting*. Upper Saddle River, NJ: Prentice Hall.

2.15 Sage, A., and C. Cuppan (2001). On the Systems Engineering and Management of Systems of Systems and Federations of Systems, *Information, Knowledge, and Systems Management*, Vol. 2, No. 4, pp. 325–345.

2.16 Maier, M. Architecting Principles for Systems of Systems, www.infoed.com/Open/PAPERS/systems.htm.

2.17 Keating, C., et al. (2003). System of Systems Engineering, *Engineering Management Journal*, Vol. 15, No. 3, September.

Chapter 3

Problems to Ponder

Building and managing complex systems implies a large number of problems that need to be solved. One's success with respect to these systems is therefore tied directly to solving a reasonable number of these problems. Many problems can be considered straightforward in that various solutions are basically common knowledge. Others require special efforts, including an ability to think outside the box. In short, the builder and manager of complex systems must also be able to solve problems, at times even imperfectly, in a world that often does not behave according to a predictable script.

The general steps in conventional problem solving are mostly well known. For example, here is a short list of mine from an earlier work [3.1]:

1. Surfacing a problem
2. Articulating the problem
3. Offering alternatives
4. Evaluating alternatives
5. Implementing a solution

In the remainder of this chapter we identify and discuss problem areas pertaining to systems, people, software, and management. Try working your way to solutions with which you are comfortable. You might also try to use the steps cited above. After you've read the entire book, you might also wish to attempt other approaches that are derived from the nine perspectives that are explored directly in Chapters 5 through 13.

Managing Complex Systems: Thinking Outside the Box, By Howard Eisner
Copyright © 2005 John Wiley & Sons, Inc.

3.1   PROBLEM AREAS: SYSTEMS

Part of building and managing systems in a better way is to understand the problems that one might be faced with in this endeavor. A brief summary of some of these problems, from a systems perspective, follows:

1. Stovepipes not integrating as expected
2. Need for system comes into question
3. Insufficient funding
4. Insufficient schedule
5. System failing during test and evaluation
6. Requirements creeping and not satisfied
7. Scope arguments with customer
8. Insufficient tool support (e.g., simulation)
9. No processes; all ad hoc
10. Earned-value problems

3.1.1   Stovepipes Not Integrating as Expected

The matter of attempting to integrate stovepipe systems was discussed briefly in Chapters 1 and 2. Although the assumption is often made that such stovepipes can be integrated in a cost-effective manner, this is not necessarily the case. Much care needs to be exercised with respect to (1) how much integration is desirable and (2) in what order the integration should take place.

3.1.2   Need for System Comes into Question

Whether in a commercial or a government setting, there are times when the need for a particular system may be questioned. An example in the government world is the national missile defense program. Over the years since it was introduced as the strategic defense initiative program, many high-level critics, especially in the Congress, have argued that there is no need for such a system. In the commercial world, there are often vice presidents who do not support the introduction of a new system, especially when they perceive that a legacy system can last for a few more years.

3.1.3   Insufficient Funding

The problem of insufficient funding is very common. Many projects, especially with respect to large complex systems, do not have enough funding allocated at the start of the program. It is assumed that additional funding can be obtained, but often it is not. Lack of assured funding also will put massive pressure on both a contractor and a customer, leading to constant controversy and inordinate amounts of time devoted to negotiations.

3.1.4 Insufficient Schedule

The issue of "insufficient" schedule is related to a presumed requirement to have a system completed and installed on a schedule that is a severe hardship for a contractor. To meet such a requirement, which may have been promised during the proposal phase, all project personnel are asked to work 10 to 12 hours a day. This can become counterproductive and lead to a direct loss of key people. The issue might be expressed in a question: How can we assure that schedules are more reasonable for all parties?

3.1.5 System Failing During Test and Evaluation

Test and evaluation comprise the last formal and extensive demonstration that a system does indeed meet its requirements. One type of test and evaluation is carried out in an operational setting. A most serious problem occurs when the system fails at this very late date. Recovering from this result is usually a very expensive and time-consuming matter, if it can be achieved at all.

3.1.6 Requirements Creeping and Not Satisfied

History tells us that creeping requirements will often lead to project failure. Further, as they are changing, they are also failing to be satisfied. Problems with requirements appear to make everyone's top 10 list of reasons why system developments fail. This is despite the fact that we know beforehand that this is a problem area. Some of the problem seems also to be related to the form of contract under which work is being performed (e.g., firm fixed price vs. cost reimbursible).

3.1.7 Scope Arguments with Customer

The matter of arguments with the customer is related to that of requirements, especially with design and study types of efforts. It may show up during the review of a report such that it is ultimately rejected by the customer, who claims that various topics were not covered. The contractor may respond by claiming that such topics were "out of scope." For fixed-price contracts, this can be a show-stopper.

3.1.8 Insufficient Tool Support

Most of the steps in engineering a system need to be carried out with the benefit of proper computer support. Of special interest in this respect are modeling and simulation tools as well as system development environments. Although much progress has been made in this and related areas, some companies, especially the smaller ones, simply lack the resources.

3.1.9 No Processes, All Ad Hoc

Although the various capability maturity models (e.g., CMMI [3.2]) have addressed matters of process, many companies are either far behind in their implementation or too small to have dealt appropriately with the matter. Ad hoc methods, especially with larger and more complex systems, are likely to break down at some point in a project and lead to great difficulties.

3.1.10   Earned-Value Problems

A classical earned-value problem may be expressed as follows: "We are 80% complete in terms of dollars spent as well as schedule, but are only 50% complete with respect to products completed and delivered." This is clearly a problem as stated, but often the earned value is not calculated in as clear a manner as this. When earned value is unclear, the likelihood of project breakdown begins to increase, often dramatically.

3.2   PROBLEM AREAS: PEOPLE

In addition to system-oriented issues and problems, many difficulties can be attributed directly to the people side of the equation. Ten factors of this type are listed below and then discussed very briefly:

1. Divisive person on team
2. Erratic communications
3. Insufficient technical expertise
4. Bad boss
5. Bad customer
6. Standard 12-hour days
7. No authority to hire and fire
8. People not sharing information
9. Information services inadequate
10. Contracts organization won't support

3.2.1   Divisive Person on Team

There are times when a divisive person shows up on a project whose members are attempting to build a team. Such a person is a threat to the formation of a true team and usually needs to be removed from the project. Project management is often slow to deal with this issue, not understanding that a dysfunctional team may be the result.

3.2.2   Erratic Communications

We are told that many of our problems are related directly to our failure to communicate, both at meetings as well as one-on-one and in written reports. Specific efforts need to be put forth to improve communications and make it a strength rather than a weakness. Project managers need to be diligent in making sure that misunderstandings are minimal or that they disappear altogether.

3.2.3 Insufficient Technical Expertise

Systems are designed and built by people, and when they lack the required technical expertise, system development is likely to be a serious problem. This deficiency, of course, will be magnified when the system is large and complex. In this scenario, special attention needs to be paid to assure the selection of competent project personnel.

3.2.4 Bad Boss

In today's world especially, where there tends to be greater job mobility as we move through the information age, people will not tolerate a bad boss for very long. Further, there is still a considerable amount of evidence that despite all the training, both formal and on-the-job, bad bosses still inhabit the industrial as well as government worlds in great numbers.

3.2.5 Bad Customer

Many people complain about having to deal with a bad customer. There are many reasons for such a complaint, ranging from a very demanding micromanaging customer to one who appears not to know what he or she is doing. Many bad customers think that contractors are the equivalent of slaves. Many do not accept responsibility in one of the key system development areas: the articulation of a coherent and correct set of requirements.

3.2.6 Standard 12-Hour Days

There are many reasons why an enterprise may resort to having its people work a 12-hour day, most of them not really justifiable, at least as a standard operating practice. Some of the problems cited above (e.g., behind schedule, overspent) will lead to 12-hour days in order to catch up. For professional personnel who do not get paid for overtime, a steady diet of 12-hour days will often lead to most team members looking for another job.

3.2.7 No Authority to Hire and Fire

Project managers or team leads need more authority to hire and fire persons who are not contributing as they should be. This is especially true for complex systems. It is also true when, as suggested above, a person is being a divisive force on a program.

3.2.8 People Not Sharing Information

Information has become more and more a means by which people create and use power. By not sharing important information, they perceive that they have an advantage over co-workers and bosses. This is the antithesis of team building, and should not be tolerated when it is discovered. At the same time, project managers need to know how to set the stage as well as how to set an example for the sharing of information.

3.2.9   Information Services Inadequate

In many enterprises, there is a chief information officer (CIO), whose job it is to make sure the organization has the information systems that it needs to conduct business. Many CIOs do not have the budget or the service orientation to make sure that the true information needs are being met. A simple example is the project cost accounting system that all project managers need to be effective at the grass-roots management level.

3.2.10   Contracts Organization Won't Support

In a manner that is analogous to the problem area above, the contracts department in any organization is supposed to support the line organization at all levels, to make sure that managers are aware of the contract provisions under which they are operating. When this transfer of information and support is not present, the organization tends to be operating at greater risk as well as reduced efficiency.

The discussions of systems and people problems above remind us of three aspects of building and managing complex systems. The first is that very specific and often predictable problems arise, despite our attempts to keep them from doing so. The second is that success often depends, in large measure, on our ability to be good problem solvers (i.e., to bring closure to these problems in acceptable ways). Finally, we believe that success in problem solving can be facilitated by being able to think outside the box. Let us now explore two other types of problems, those pertaining to software as well as overall management issues.

3.3   PROBLEM AREAS: SOFTWARE

Following are some problem areas that may show themselves with respect to software development:

1. Not able to confirm build completion
2. Cannot integrate software from disparate sources
3. Cannot integrate COTS software
4. Cost and schedule estimates inaccurate
5. Widely disparate programmer productivity
6. Cannot evaluate and mitigate software risks
7. Poor software architecting process

8. Do not know how to deal with software warranties
9. Cannot compete with outsourced software development
10. Cannot satisfy a critical customer requirement for software performance

As in previous sections, a brief discussion of each problem will be presented, generally stopping short of describing a "solution." It is suggested that the reader look at each of the above and at the same time write down some thoughts about a potential solution. After reading all the chapters that present ways to think outside the box, you may wish to go back to these problem areas to see if different solutions come to mind.

3.3.1   Cannot Confirm Build Completion

Although software is usually designed top down, it is generally built bottom up. A *build* is a discrete and identifiable chunk of software, usually made up of several software units. Even though all of the units that go into a build may pass unit testing, we cannot be sure that we have completed a build until its units have been integrated and the overall build has passed the appropriate testing. Further, even if a build passes integration and test, it must still be integrated with other builds. All of this creates uncertainties as to the actual completion status of the builds.

3.3.2   Cannot Integrate Software from Disparate Sources

The next problem often materializes when a system is designed with significant software reuse. Even though an overall system plan calls for reuse, trying to integrate software from a variety of sources can fail. Software engineers are well aware of this and sometimes show a disinclination to sign up for software reuse. Even greater difficulties are experienced when low bids are based on large amounts of software reuse which cannot be realized during system implementation.

3.3.3   Cannot Integrate COTS Software

This can be considered a special case of the above, and represents one of the most difficult software integration problems and issues. For example, it is one thing to incorporate Microsoft Office into an overall system but is quite another to attempt to integrate it with other functions that are part of the system.

3.3.4   Cost and Schedule Estimates Inaccurate

Cost and schedule inaccuracies are usually not discovered until some actual schedules and costs have been experienced. It is then that we see the differences between initial estimates and the realities of projects and programs. This type of problem continues to show itself even though we use well-known methods for making such estimates, such as COCOMO [3.3] and function point analysis [3.4].

3.3.5 Wildly Disparate Programmer Productivity

We have experienced the fact that a group of programmers at the same pay grade can exhibit more than one-order-of-magnitude difference in productivity. This can create considerable chaos in both planning and implementation. This problem can often be exacerbated by the fact that our more senior and productive software programmers are asked to do the planning. They use themselves as a benchmark, leading to optimistic estimates that cannot be met by a mix of personnel.

3.3.6 Cannot Evaluate and Mitigate Software Risks

Software risks may be placed in any one or more of the following categories: (1) schedule risk, (2) cost risk, (3) performance risk, and (4) administrative risk. The most difficult of these may well be performance risk, and if the software fails to meet technical performance specifications, it is likely that problems with schedule and cost will soon follow. Also, this problem usually becomes more severe when we are dealing with real-time systems.

3.3.7 Poor System Architecting Process

The design of a large-scale, complex software system requires many skills, one of the most important being a solid and repeatable process for developing a software architecture. This, in turn, is dependent on notions of software decomposition, which can be quite variable. Clearly, working on a software system with a poor or incomplete architecture is likely to lead us down the road to failure.

3.3.8 Do Not Know How to Deal with Software Warranties

In today's world, customers are likely to specify that the software delivered be free of latent and patent defects. If not able to estimate the levels of such defects, software producers are unable to produce or offer meaningful warranties. Systems with software problems, delivered under a firm-fixed-price (FFP) contract, can result in massive losses to the developer and frustration for the customer.

3.3.9 Cannot Compete with Outsourcing

We are seeing the realities of large amounts of software being outsourced to foreign developers at a price considerably lower than that experienced at home. The results of this practice have been quite respectable, despite cries that such a process is not workable from a technical and management point of view. If you are trying to be competitive with your software, what approach should you take?

3.3.10 Cannot Satisfy a Critical Requirement

A typical area in which a problem might be experienced regarding satisfying a critical requirement is software run time or response. The critical dimension is time, and the software just won't run fast enough as defined by the system requirements and specifications. Both the success of systems and the careers of software engineers can hinge on a single run-time requirement in the contractual statement of work.

3.4 PROBLEM AREAS: MANAGEMENT

The final set of problem areas in this chapter pertains to the field of management, from the corporate level all the way down to the project level. Ten such areas are cited below:

1. Poor strategic planning and decision making
2. Insufficient yearly growth
3. Quality of services and products inadequate
4. Weak in research and development (R&D) investments
5. Operations not profitable
6. A follower in all lines of business
7. Pattern of losses to competitors
8. Poor management at project level
9. Enterprise not sufficiently agile
10. Poor in Senge's [3.5] five disciplines

3.4.1 Poor Strategic Planning and Decision Making

Poor strategic planning and decision making applies to the nominal leaders in an organization, dealing with matters of overall strategy and its implementation. The selection of the vision and "correct path" for an enterprise affects projects that the organization will undertake over the long term. For example, when Xerox decided to bet everything on xerography, the key focus was how to try to make that bet work, overshadowing just about everything else the company was doing. From another perspective, when Wang Laboratories decided, overtly or otherwise, not to pay much attention to how the world was changing when it was a dominant player in desktop word processing, it set the stage for the eventual bankruptcy of the company.

3.4.2 Insufficient Yearly Growth

The top-level growth goal for an enterprise can have an immediate impact on the projects it will undertake. A high yearly growth goal, for example, can put pressure on an enterprise such that month-to-month increases in revenue become critically important. Everything else is subordinate, and immediate results are emphasized. Such a perspective will stifle other activities, such as R&D investments, which tend to be focused on a longer time frame. This is a classical problem that many public companies experience as they perceive that quarter-to-quarter results signal the financial community, which, in turn, helps to determine the price of the company's stock.

3.4.3   Quality of Services and Products Inadequate

Poor quality of services and products can be sufficient to bring down an enterprise. As a general principle, a policy of continuous improvement can serve to turn such a situation around. The field of total quality management is supported by an extensive literature regarding how to deal with this type of problem (see, e.g., [3.6, 3.7]).

3.4.4   Weak in R&D Investments

As alluded to above, making little-to-none or misguided investment in research and development can lead to disasters, especially for a company in the high-tech arena. Making these "bets" is a difficult matter, especially for midsized enterprises that do not have a lot of discretionary dollars to spend. There is no way to ensure success, but there are many guidelines available as to how to carry out a sensible program for investment in a company's future.

3.4.5   Operations Not Profitable

A negative bottom line, especially over several years, is a problem that requires immediate attention. It is literally amazing to see many companies operate in the red, yet continue to survive. One such company is Amazon.com, which seems to have successfully demonstrated other variables can overcome red ink at the bottom line. It would appear that their behavior with respect to profitability is unique and not recommended as a mode of operation to be emulated.

3.4.6   A Follower in All Lines of Business

Companies tend to define and follow various lines of business (LOBs) that help them establish who they are and their selected areas of focus. In this respect one can recall the Jack Welch criterion for GE: they wanted to be either number one or number two in all of their important LOBs [3.8]. Many attribute GE's great success to maintaining this type of perspective as a way to assure that they continue to invest in and build market leaders.

3.4.7   Pattern of Losses to Competitors

Very few companies are in a position to dominate their industry. A distinct pattern of losses to competitors is obviously a signal that something is wrong. Diagnosing this type of problem can be difficult, especially when remedial actions have been taken to change the situation. An example is when a government contractor continues to cut costs and thereby reduce its bid prices with respect to both overhead and general and administrative expenses. Perhaps the true problem lies elsewhere.

3.4.8   Poor Management at Project Level

Many companies simply assume that grass-roots project managers know how to manage small projects in an effective manner. This is a dangerous assumption that can lead to disaster, especially on firm-fixed-price contracts. A learning organization, as defined by Senge [3.5] and others, will provide a blueprint for companies trying to avoid this type of problem.

3.4.9   Enterprise Not Sufficiently Agile

There is considerable pressure for companies to learn how to "shorten the time line"; that is, there is a premium for being able to operate with reduced schedules, especially for companies that are involved in national defense and security matters. This is subject to a ripple effect, because large prime contractors tackling the construction of complex systems will insist that all levels of subcontractors be able to respond quickly with superior products and services.

3.4.10   Poor in Senge's Five Disciplines

Peter Senge has defined the following five areas as critical to the success of an enterprise [3.5]:

1. Personal mastery
2. Mental models
3. Building shared vision
4. Team learning
5. Systems thinking

The latter, established as the fifth discipline, is an important area of consideration in this book, especially in Chapter 13.

3.5   SUMMARY

In this chapter we have posed 40 problems that can show themselves as we design, build, and manage complex systems. Clearly, even such a long list is not exhaustive. However, its purpose is to suggest three perspectives:

1. Complex systems carry with them a host of difficult problems.
2. Success with these systems requires that we learn how to be better problem solvers.
3. Better problem solving may, in turn, require us to be able to think outside the box.

The reason for generally not suggesting solutions to these 40 problems is to leave them as areas for the reader to think about both *before* and *after* reading this book in its entirety. As noted above, there are times when an out-of-the-box approach gives us just what we need to find an answer to what has appeared to be an insoluble or intractable problem.

In Chapter 4, we turn our attention to the inventive mind. Although we do not expect to duplicate the genius of a Leonardo da Vinci or an Edison or a Newton, a brief overview of what these larger-than-life but still mortal people were able to achieve should help us as we seek more powerful solutions. Although we cannot hope to attain their penetrating levels of reasoning, just looking at what they have produced can be nothing less than inspiring.

REFERENCES

3.1 Eisner, H. (2000). *Reengineering Yourself and Your Company: From Engineer to Manager to Leader.* Norwood, MA: Artech House.

3.2 CMMI is the integrated capability maturity model; information may be found at www.sei.cmu.edu.

3.3 Boehm, B., et. al. (2000). *Software Cost Estimation with COCOMO II.* Upper Saddle River, NJ: Prentice Hall.

3.4 Jones, C. (1997). *Applied Software Measurement*, 2nd ed. New York: McGraw-Hill.

3.5 Senge, P. (1990). *The Fifth Discipline.* New York: Doubleday Currency.

3.6 Walton, M. (1986). *The Deming Management Method.* New York: Perigee Books.

3.7 Feigenbaum, A. (1991). *Total Quality Control*, 3rd ed. New York: McGraw-Hill.

3.8 Welch, J. (2001). *Jack—Straight from the Gut.* New York: Warner Books.

Chapter **4**

The Inventive Mind

This chapter focuses on several additional examples with respect to extraordinary thinking. As a prelude to these examples, an overview of the basic idea of thinking outside the box can be illustrated by reference to Figure 4.1. Here we see the *normal* or *Gaussian distribution*, otherwise known as the *bell-shaped curve*. We can break this curve into four regions, roughly representing a person's ability to find an appropriate solution to a knotty problem. These regions are four cases in which:

1. An extraordinary solution is found (far right of the distribution) (region 1)
2. A possible solution is found (to the right of the midpoint or mean) (region 2)
3. A poor solution is found (just to the left of the midpoint or mean) (region 3)
4. An unacceptable (or no) solution is found (far left of the distribution) (region 4)

We are going to place a few *notional* numbers on these four regions. For purposes of thinking about the concept, consider these regions as follows:

- Region 1: contains about 5% of the distribution
- Region 2: contains about 45% of the distribution
- Region 3: contains about 45% of the distribution
- Region 4: contains about 5% of the distribution

Managing Complex Systems: Thinking Outside the Box,   By Howard Eisner
Copyright © 2005 John Wiley & Sons, Inc.

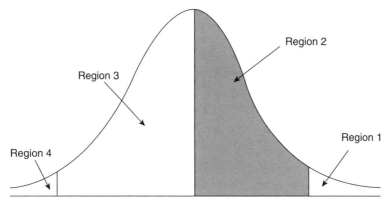

Figure 4.1 Normal distribution and notional regions.

Region 1 is meant to represent true thinking outside the box, with only about 5% of our problem solvers developing an extraordinary solution. As time goes on, and as we build learning organizations [4.1], we would hope that this number would rise to as much as 10 to 15%. What this overview says is:

1. Currently, a rather limited number of people (perhaps the top 5%) are thinking outside the box.
2. We would hope that with training and practice we can change this to about the top 10 to 15%.

Can the latter goal be achieved or even exceeded? Only time and some data as we move on down the road will help answer this question.

The *conjecture* or *assumption* that true thinking outside the box is related to about a top 5% also implies that certain problems will be faced by such thinkers. What they are, and how to cope with them, will be reserved for additional discussion in a later chapter. At this time, let us explore other examples of special ways of thinking from a number of people who clearly had inventive minds. These particular people probably operated within a fractional part of the top 1% and had other skills (e.g., determination) that contributed to their success. For our current purposes, they serve as models of both extraordinary thinking and achievement that the rest of us can attempt to emulate.

4.1 MANAGEMENT THINKING

4.1.1 Drucker's Management by Objectives

Some years ago, Peter Drucker introduced one of the clearest and most successful cases of out-of-the-box thinking that had ever been offered in the world of management. He set forth his ideas on management by objectives (MBO), which required a focus on achieving a set of objectives that were agreed upon in advance by the appropriate levels of management. Drucker, the grand master of management consultation, argued that by focusing on end results, you allow employees to create their own means to achieve the results. This avoids micromanagement and encourages the development of new processes that represent improvements in the old processes. Beyond this "theory," it is fair to say that the adoption of MBO principles and practices has been extremely widespread and enormously successful. Drucker can certainly be regarded as one of our best out-of-the-box thinkers, and his ideas range far and deep in many aspects of keeping enterprises efficient and effective [4.2–4.4].

4.1.2 Xerox's Technology and Alliances

The early days of the Xerox Corporation are marked by distinctive and almost singular out-of-the-box thinking. As a relatively small company in Rochester, New York, the Haloid company (predecessor of Xerox) contemplated its future in rather uncertain terms. As a manufacturer largely of photographic paper, Haloid had considerable difficulty seeing how it could be competitive over the long haul. Led by its president, Joe Wilson, and its corporate attorney, Sol Linowitz, Haloid decided to invest heavily in xerography technology, specifically in the process invented by Chester Carlson at the Battelle Memorial Institute [4.5, 4.6]. Indeed, they agreed to "bet the company," with the prospect of creating a new industry. In a very real sense, they were so far out of the box that both IBM and RCA turned down strategic partnering overtures on the grounds, apparently, that xerography would never take hold. Or perhaps they wanted to "go it alone."

Changing its name to Xerox, the company went further outside the box and sought alliances outside the United States. This high-risk and somewhat revolutionary strategy resulted in alliances with both Rank (U.K.) and Fuji (Japan), giving the company much needed financial backing as well as entry points in international markets. As a small company, early Xerox needed this type of support to be able to bring this new technology to worldwide markets quickly and efficiently. Literally, a new industry was being created under the noses of U.S. leaders such as IBM, GE, RCA, and others. Who would have believed that this could happen?

As the company flourished, they went out of the box another time by setting up the Xerox Palo Alto Research Center (PARC) [4.7]. This storied center had a very broad charter: to explore technologies that might be important to the company just as xerography had turned out to be a vital part of the company's success. Xerox PARC indeed came up with several good ideas. They were ahead of their time in such areas as the use of a mouse, on-screen human–machine interfacing, computer architectures, and others. Unfortunately, Xerox headquarters failed to bring these new ideas into workable products. Instead, they invested considerable resources in other areas (e.g., real estate) and began to lose ground in at least two dimensions:

1. They failed to commercialize the excellent technologies brought forth by Xerox PARC.
2. They failed by investing in domains about which they knew little and in which they lost vast amounts of money.

So Xerox could not sustain their leading-edge thinking and started down a bureaucratic road that led to many years of great difficulty, all of this despite the fact that their technologies contributed in a major way to the success of other enterprises.

4.1.3 Deming and the Pursuit of Quality

Dr. Deming, as those in industry came to know him, was ahead of his time with out-of-the-box thinking and as a result was not recognized as a guru of sorts until later in his career. Trained as a statistician, W. Edwards Deming was a deep thinker about quality control and management processes that could be applied to the building of just about all products, of all shapes and sizes. Not finding an environment in the United States that was conducive to his ideas, he went off to Japan to consult for a variety of their industries. He was more than accepted there, and rose to almost legendary status as he helped the Japanese significantly improve their products as well as the processes that produced them [4.8].

Students of Deming are familiar with his *Fourteen Points*, which are:

1. Create a constancy of purpose to improve both products and services.
2. Adopt the new philosophy (from a management perspective).
3. Stop relying on mass inspections as a means of achieving quality.
4. Cease the practice of awarding business on the basis of lowest cost on an item; instead, attempt to minimize total cost.
5. Institute the practice of continuous improvement with respect to production and service.
6. Assure effective training on the job.
7. Focus on ways and means of encouraging leadership.
8. Drive out fear at all levels in the organization.
9. Break down barriers that keep departments from working together constructively.
10. Eliminate slogans that appear to take the place of the real work of improving quality.
11. Do away with work quotas on the factory floor.
12. Eliminate barriers to pride with respect to all work.

13. Assure that employees are enhancing their skills through education and self-improvement.
14. The transformation represented by the above is everyone's task.

These points provide a compact and somewhat simplified version of Deming's overall approach but nonetheless were ultimately embraced by business and industry in his homeland. Deming lectured extensively in the States, expounding these 14 points and giving examples that were drawn from his firsthand experiences around the world. Deming became the leading guru of total quality management and its derivative ideas. Here is a case where initial out-of-the-box thinking persisted so long that it eventually transmuted into the conventional wisdom of the day. As such, it has yet to be challenged successfully or replaced by a new structure for producing high-quality products.

4.1.4   Augustine's Laws and Travels

I met Norman Augustine in the late 1980s at his office in Bethesda, Maryland. I was gathering material for my book on computer-aided systems engineering and I wanted to include some of Augustine's laws [4.9] in the book. I asked for his permission to do so, and he gave it to me, with one proviso—that I donate an autographed copy of my book to his company. I quickly agreed, and followed up a couple of years later when the book was published. We also talked a bit about engineering in general and systems engineering in particular. These meetings clearly demonstrated to me how knowledgable as well as personable he was.

Norman Augustine had a lot to do with establishing how to survive in the post–cold war period when defense expenditures were phasing down, catching a lot of military-oriented companies by surprise. His thesis was that a lot of companies were going to disappear and that the way to try to survive was to begin an aggressive acquisition program, coupled with increased emphasis on how to achieve internal organic growth. He set this in motion as the key executive at Martin Marietta and was the prime mover in bringing about the merger with Lockheed. This was not an easy path, but he carried it forward with his typical honesty and grace. Today, Lockheed Martin is one of the largest and most respected companies in the world. A lot of credit for its evolution goes to Norman Augustine.

Mr. Augustine (who has 13 honorary doctoral degrees) retired from Lockheed Martin in 1997 and went off to teach at Princeton, pursuing a new and broader career as a teacher, writer, and consultant. His later book [4.10], which dealt with many of his business experiences (as well as his early desire to be a forest ranger), has many interesting observations about such topics as:

1. What it is that makes people leaders
2. Six stages of crises
3. Reengineering
4. A 12-step process for negotiating

5. Corporate diversification
6. Mergers and acquisitions

Norman Augustine continues to make important contributions to how we think about government and industry and has set high standards for those who have come after him.

4.1.5 Jack Welch and GE

Jack Welch joined General Electric (GE) in 1960 with a recent doctorate in engineering. Twenty-one years later (1981) he was installed as CEO and chairman of this behemoth company. Twenty-years later, he retired, having made an indelible imprint on the company, building its market capital by more than $450 billion (!). He was also known as "neutron Jack," as a consequence of his tendency to downsize when he thought it was necessary. Indeed, his strategy was called by many a "fix, sell, or close" approach, as he appeared to be so matter of fact about what had to be done. In his search for excellence, he wanted the company to be either first or second in all its various businesses—so much for setting simple but powerful guidelines for what to do in an enterprise.

By his own admission [4.11], Welch was a basher of bureaucracy par excellence. In a company with the size and inertia of a GE, there was lots of bureaucracy to bash, and nobody did it better and more effectively. Welch points to four major initiatives that he took on in the 1990s to continue to grow the company:

1. Globalization
2. Services
3. Six Sigma
4. E-business

He approached every one of these with a powerful focus and passionate commitment to make them successful. If Deming espoused "constancy of purpose," Welch brought that idea to the level of both a fine art and a science. If he thought, and he did, that these were the four initiatives that would bring GE into the twenty-first century in the best possible way, nobody was going to stand in his way. If you were not able to adjust, to make the changes necessary to move on, your tenure was going to be short. Indeed, when he left GE he felt that he had made GE an enterprise that had learned to love change. In doing so, he showed an entire generation of managers how an old-school, hands-on engineer could be one of the most successful technical business leaders of our time.

4.1.6 Harold Geneen and ITT

One might say that Harold Geneen was to ITT as Jack Welch was to GE. Both were enormously successful in building a solidly profitable enterprise. Both placed enormous stock in the people side of the equation and prided themselves in making excellent personnel choices for the benefit of the company. Both were devoted to bashing the bureaucracy and coming down to simple, straightforward, honest talk. Both were committed to looking for the unadorned facts in all situations and insisting on numbers to back up or represent these facts. Both were very tough-minded and disciplined, completely devoted to obtaining results that could be observed by all. Both used acquisitions to grow the enterprise, and with extra-ordinary success. However, Geneen preceded Welch by close to 20 years. In that sense, Geneen could have been an excellent model for Welch to take into account as he took the helm at GE.

Harold Geneen started as CEO and chairman of ITT in 1959 when the company had sales of $765.6 million and profits of some $29 million [4.12]. When he left the company in 1977, sales were $16.7 billion (!) and profits were $562 million. ITT had bought or merged some 350 businesses in 80 countries, resulting in about 250 profit centers. At its peak, ITT was the ninth largest industrial firm on the Fortune 500 list. The growth of the company included 58 consecutive quarters of improving financial results.

Here are 10 areas of emphasis that can be gleaned from a careful review of Geneen's own discussion of how he managed [4.12]:

1. A company is judged by one important criterion: performance.
2. Quality of earnings was one critical measure of performance.
3. All managers had to think of ITT as one company and behave accordingly.
4. A free flow of information was crucial to personal and corporate success.
5. Monthly operating reports had to contain (a) facts and (b) red flags to identify true problem areas as well as mitigating plans and actions.
6. Face-to-face meetings were standard operating procedure.
7. Deep and solid communications were critical, resulting in lots of meetings.
8. Management must manage and must achieve agreed-upon results.
9. An indicator of success was a positive "emotional attitude."
10. You had to play by the rules at all times, but that did not mean you had to *think* by the rules.

Item 10 can be thought of as Geneen's invitation, in his day, to think outside the box. He certainly was able to do exactly that, and he encouraged his people to do the same.

4.2 SCIENTIFIC AND TECHNICAL THINKING

The discussion above highlighted some of the achievements and principles of a selected group of industry leaders. These people clearly thought for themselves and set new directions in the often confusing world of large and complex systems.

TABLE 4.1   Selected Scientific and Technical Contributors

Contributor	Time Period	Known for:
Descartes	Born in La Haye, France, in 1596	Logic; mathematics; metaphysics
Edison	Second half of nineteenth century	Incandescent light; phonograph; electric heat; light and power systems
Einstein	First half of twentieth century	Theories of relativity; space–time; energy–mass relationship
Faraday	Nineteenth century	First electric motor; generator; transformer
Franklin	Eighteenth century	Helped draft Declaration of Independence; stove; bifocals; kite and electricity
Galileo	Born in Pisa, Italy, in 1564	Sun as center of solar system; pendulum; superior telescope
Leonardo da Vinci	From apprenticeship in Florence in 1469 to his death in France in 1519	Art; aerodynamics; hydrodynamics; weapons
Maxwell	Second half of nineteenth century	Electromagnetic equations; color photography
Newton	Mid to late seventeenth century	Laws of motion and gravitation; calculus; optics
Salk	Twentieth century	Vaccine for polio
Watson and Crick	Twentieth century	DNA double-helix molecular model
Wright Brothers	Early part of twentieth century; flight at Kittyhawk, NC, in 1903	First heavier-than-air aircraft design and flight

Examining carefully what they said and what they did will serve the rest of us in good stead as we tackle current and future problems.

There is another group of people with very inventive minds that have made major contributions to our society: scientists and engineers who have forged new pathways over long periods of time. Our technological advances are based on their good work, and we owe our standard of living as well as our quality of life to that work. Table 4.1 provides a very short list of some of these advanced thinkers and the principal achievements for which they are known.

It is not possible to be sure about what several of these and other inventive people thought in detail and how their thinking evolved. Suffice it to say that they all questioned conventional wisdom, and that allowed them to move ahead into previously uncharted territories. They probably asked themselves the key question: Is there another way to think about this problem, another way to look at it? Many of us know how to pose this question but find that going beyond the question is replete with difficulties that reflect our own limitations. But our goal in this book is not to find out how to seek the one-in-a-million mind, but rather, to bring a few (actually, nine) new perspectives to the table that have a good chance of leading to better solutions in the challenging but down-to-earth world of building and managing systems. To that end, the following short hypothetical example illustrates how the better idea can lead to positive and possibly surprising results in a corporate environment.

4.3   NOT EVERYONE IS AN EINSTEIN (OR NEEDS TO BE)

XYZ Corporation had experienced growth in revenues and profits of approximately 8%, 3%, and 1% over the past three years. The president found this unacceptable and convened a series of meetings of his vice presidents to examine this important issue and what to do about it. These meetings went on for about three weeks, and a consensus view was starting to form that more substantial levels of growth could be achieved by starting an acquisition program. By means of the revenues and profits of the companies acquired, acceptable growth levels could be assured.

The president was pleased and so were the vice presidents—or so it seemed. It was then that Jennifer, the company VP for administration, got up to speak. She pointed out that the company had never done an acquisition, which carried with it at least two very large challenges. One was to service the debt that would be incurred by means of the proposed approach to acquisition, and the other was to manage the acquired company in such a way as to assure that all came out well, including the desired financial results. She pointed out some of the pitfalls she had encountered when she was at another company, all of which suggested the high risk of going with the acquisition approach. "It appears to be the easy solution," she said, "by buying our way out of our problems." "But our real problem," she continued, "is that we have lost our ability to grow from within." She proceeded to lay out a well-conceived plan for internal organic growth that, she claimed, (1) was a critical need of the company, (2) had a greater chance of success, (3) had considerably lower cost, and (4) would lead to more value for the current stockholders, including themselves. As an enlightened management team they were able to rally around Jennifer's out-of-the-box thinking. They scuttled the acquisition plan, at least for the time being, and adopted her plan for continuous internal improvements. The company experienced healthy growth from within and benefited greatly from becoming a continuously improving and *learning organization* [4.1].

The point of the example is simply to show that a better idea or solution does not need to be earthshaking or worthy of a Nobel prize. Jennifer could see that the conventional wisdom accepted by her colleagues was risky, and another approach appeared to her to be more feasible. To her credit, she had both a better idea and the courage of her conviction. By thinking outside the box and doing some extra homework, she was able to carry the day, leading to beneficial results for the company. That's sufficient as a success story for most of us.

4.4 THE NEXT NINE CHAPTERS

The next nine chapters define perspectives for thinking outside the box. These are described in specific terms, with many examples. Every one of these perspectives has, at one time or another, been part of my life. They are a distillation of approaches that have worked in varying degrees in various situations. It is hoped that some of them will work for the reader as he or she approaches the problems of building and managing complex systems, and problem solving in general.

REFERENCES

4.1 Senge, P. (1990). *The Fifth Discipline*. New York: Doubleday Currency.

4.2 Drucker, P. (1964). *Concept of the Corporation*. New York: Mentor Book.

4.3 Drucker, P. (1966). *The Effective Executive*. New York: Harper & Row.

4.4 Drucker, P. (1985). *Innovation and Entrepreneurship*. New York: Harper & Row, Perennial Library.

4.5 Linowitz, S. (1985). *The Making of a Public Man*. Boston: Little, Brown.

4.6 Owen, D. (2004). *Copies in Seconds*. New York: Simon & Schuster.

4.7 Cringely, R. (1996). *Accidental Empires*. New York: HarperCollins.

4.8 Walton, M. (1986). *The Deming Management Method*. New York: Perigee Books.

4.9 Augustine, N. (1982). *Augustine's Laws*. New York: American Institute of Aeronautics and Astronautics.

4.10 Augustine, N. (1998). *Augustine's Travels*. New York: American Management Association.

4.11 Welch, J. (2001). *Jack—Straight from the Gut*. New York: Warner Books.

4.12 Geneen, H. (with A. Moscow) (1984). *Managing*. New York: Avon Books.

Chapter 5

Perspective 1: Broaden and Generalize

This first perspective is extensive in its application areas, relating to businesses, individuals, and teams that are building and managing systems. A good example of what is meant by "broaden and generalize" is the rather well-known case of strategic planning by several railroads around the beginning of the twentieth century. As the story goes, railroads were asking themselves the question: What business are we in? Apparently, the answer was: We're in the railroading business. And with this answer, they kept themselves from thinking more broadly. For example, they could have said that they were in the transportation business. By not doing so, they kept themselves from seeing new opportunities in other forms of transportation, such as air, maritime, urban mass transit, and automobile. Thus, a single restrictive perspective set the course for the future of many, if not all, of these important railroading businesses. Broader views set the stage for, but do not guarantee, the successful mining of new opportunities.

The example above pertains to the strategic mindsets of businesses, and we can see other examples of both successes and failures in that domain. Here are some examples of such strategic perspectives, several of which are discussed in greater detail in this and other chapters:

1. IBM seeing mainframe computer hardware as their key focus
2. IBM undervaluing the importance of software
3. Microsoft seeing the value of software as a business, and shifting into operating systems when an opportunity with IBM presented itself

Managing Complex Systems: Thinking Outside the Box, By Howard Eisner
Copyright © 2005 John Wiley & Sons, Inc.

4. Wang Labs thinking that their lock on word-processing hardware and software would hold "forever"

5. Haloid betting everything on a copying process and transforming themselves into the Xerox Corporation

6. Netscape knowing when to sell the company to AOL

As cited earlier, the notion of *broaden and generalize* may also be applied to a person's views of himself or herself. Thinking about one's career in narrow terms might lead someone to conclude that he or she is a computer programmer instead of a computer scientist or engineer. Another example might be a self-perception as a database administrator rather than as a data mining scientist or knowledge engineer. Many people have made significant transitions as a consequence of their viewpoint as well as their recognition that a career is a lifelong experience that usually provides new opportunities for change of direction as well as growth along the way.

Finally, a team that is building or managing a new complex system can take a narrow view or a broad view. Both have advantages and disadvantages, the significance of which may vary widely in different situations. For example, a classical view of designing a system is to try to build it so that it satisfies all the requirements at a minimum cost to the customer. Although this approach may be successful under a broad range of circumstances, it may also be too narrow a focus for yet other situations. This low-end design approach may well neglect the fact that many customers are willing to pay more for a system that clearly exceeds the stated requirements. This is sometimes called a *best-value* approach. Yet a third notion is that some customers want the very best system, almost without much concern for its cost. Several high-end weapon systems (e.g., the joint strike fighter) might fall into this category. We use this specific domain as our first detailed example of how a broaden and generalize approach might work.

5.1 ARCHITECTING A COMPLEX SYSTEM

We will now examine a specific method for architecting a complex system that is based on the cost-effectiveness notions cited above [5.1]. This method also represents a concrete example of how the idea of broadening and generalizing can be applied to a real-world problem.

We ultimately wish to find a cost-effective architecture that is a balanced solution to a customer's problem. In doing so, we broaden our perspective, from the beginning, to three architectures rather than one. Referring to Figure 5.1, we recognize three system architectural approaches, which will produce (1) a low-cost system, (2) a high-performance or effectiveness system, and (3) a "knee-of-the-curve" system. This broadening allows us to see the possibilities that would not have been evident had we simply focused on a single approach. Single-approach thinking is inside-the-box thinking. Multiple-approach thinking takes us outside the box. The explicit consideration of alternatives is also part of what might be called the systems approach [5.1], which is examined in much greater detail in Chapter 13.

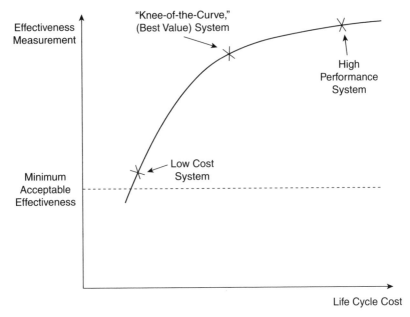

Figure 5.1 Cost-effectiveness domains for alternative systems.

The architecting process then proceeds with a functional decomposition of the system in question, leading to a synthesis matrix that defines, for all three architectures, alternative design approaches to implementing the functions and their subfunctions. This crucial step reveals the specific ways in which the functions are to be instantiated.

The next step is the formal evaluation of the three (or more, if necessary) alternative architectures. This is carried out by means of an evaluation framework that scores each of the alternative architectures against a set of evaluation criteria, which are weighted and normalized to a unity scale. The weighted scores measure the effectiveness of the alternative in question. The next step is to compute the life-cycle costs of the alternatives and then plot the results on a cost (x-axis) vs. effectiveness (y-axis) graph, as illustrated in Figure 5.1.

The architecting process described above has been developed, tested, validated, and documented empirically over many years [5.2]. It works for complex systems as well as systems of systems (see Section 5.3). We state again that at its core is the notion of broadening, especially with respect to expanding one's view to at least three alternative architectures in search ultimately of the most cost-effective architecture in regard to customer needs and constraints.

Broadening and generalizing, the first approach recommended in this book for thinking outside the box, will normally require more work than not doing so. Lifting a set of restrictive "blinders" opens up one's field of view, with more considerations now visible. Such is life in a world in which one is purposefully attempting to see what might be outside the box. The presumption is that the additional work is likely to pay handsome dividends a reasonable portion of the time. This author believes that such a presumption is supportable from real-world experience. After all, very few people have been damaged from a little more work, especially of the mental variety. Less work leading to the wrong solution is certainly not the correct approach.

5.2 USING FUNCTIONAL DECOMPOSITION TO EXPLORE STRATEGIES

We noted in Section 5.1 the role played by functional decomposition as a step in the architecting process. This same technique can be used to broaden (or sharpen) one's views of the business strategies that might be employed. Let us start with a company that builds radar systems to be used for air traffic control (ATC) purposes. Such a system might be employed in a terminal area for aircraft tracking. The shaded square in Figure 5.2 represents today's business in the area of terminal radars.

When we broaden our perspective one step, the functional decomposition diagram (FDD) shown in the figure suggests that we might explore another type of distance radar: one that pertains to the en route system. Moving up in the diagram from there leads us to the broader surveillance function and the airport surveillance detection equipment. The surveillance function is itself only one of many ATC functions, which include communications, navigation, landing support, and other systems. This can be broadened further by noting that there are other types of traffic control systems: for example, marine traffic control systems, which might be used in congested harbors. Having "discovered" that class of systems by broadening our view, we are now in a position to explore new strategies that involve any of the decomposed blocks in the figure. Some may be of interest and others not.

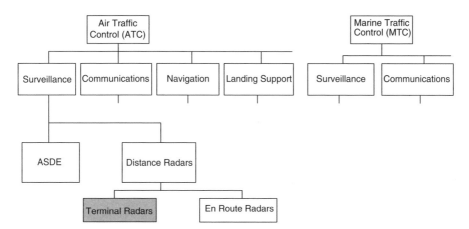

Figure 5.2 Partial decomposition of air traffic control function.

But the formal FDD approach places in evidence an array of possibilities that could ultimately lead to a new strategy and new opportunities for successful evolution of the enterprise.

5.3 SYSTEM OF SYSTEMS ENGINEERING

Another generalization that pertains to systems is moving from systems engineering to systems of systems engineering (SoSEs), referred to initially in Chapter 2. This broader view encourages us to look more deeply at (1) the differences between systems engineering and SoSE, and (2) changes in our approach given that we are moving from more conventional systems engineering to SoSE. Assume that we are building a system of systems such that each system is, or was, being constructed under the principles of the systems engineering process [5.1, 5.3, 5.4]. What is different as we attempt to integrate the systems by means of an SoSE approach?

A taxonomy for the SoSE approach suggested by me together with my co-authors focuses on integration and transition activities [5.1, 5.5]:

A. Integration Management

A.1   Scheduling
A.2   Budgeting/costing
A.3   Configuration management
A.4   Documentation

B. Integration Engineering

B.1   Requirements
B.2   Interfaces
B.3   Interoperability
B.4   Impacts
B.5   Testing
B.6   Software verification and validation
B.7   Architecture development

C. Transition Management

C.1   Transition planning
C.2   Operations Assurance
C.3   Logistics planning
C.4   Preplanned product improvement (P3I)

Another perspective regarding the broader topic of SoSE was set forth by C. Keating and several of his colleagues [5.6]. Considerations of particular interest include the following:

1. There are several problems in moving forward with SoSE.
2. These problems, if not addressed, will remain as shortcomings.
3. There are significant differences between systems engineering and SoSE.
4. A new perspective for SoSE is proposed.
5. A research agenda and set of directions for SoSE are developed.
6. The implications of SoSE for systems engineering practitioners are outlined.
7. Four implications are noted that constitute a final overview.

One "bottom line" is that SoSE will be an evolution of systems engineering and should not be revolutionary in its approach.

Yet another view was presented by M. Maier as he considered the matter of architecting systems of systems [5.7]. Some of the key points made by Maier included:

1. Systems of systems may also appropriately be called *collaborative systems* or *federated systems*.
2. There are two main criteria for the term *system of systems*: that there is both operational and managerial independence of the subordinate systems.
3. The central features defining and describing the systems of systems architecture is the communications between the systems, in particular the set of standards related to that intrasystem communications.

Sage and Cuppan explored the systems engineering and management of systems of systems and federations of systems [5.8]. Some of their conclusions were:

1. Systems of systems or federations of systems exhibit the behavior of complex adaptive systems.
2. These systems can be viewed in the context of what the authors identify as a *new federalism*.
3. Evolutionary acquisition plays an important role in these types of systems.

We note that through the processes of broadening and generalizing we are able to discover aspects of the problem that might not have been seen or addressed otherwise.

5.4 IBM's VIEW OF THE WORLD

We are not in a position to cover more than a few aspects of what might be called IBM's overall strategy, which has, of course, changed over the years. One point of departure would be some of the views of Thomas Watson, Jr., who brought IBM into its position of major prominence. Watson was the driving force behind IBM's serious commitment to computers and the consequent leadership position that the company took in that field. It was under Watson that IBM broadened its view of the business from primarily punch-card tabulators and typewriters to large-scale computers. Thinking and operating from this broader perspective allowed IBM to achieve its position of preeminence in the computer field for many years. His large investment in System/360 supported that position and contributed to enormous increases in IBM's overall value in the market in the 1960s and 1970s.

After Watson's heart attack and retirement in 1971, the company continued to do well and to broaden its approach to the marketplace. By the end of that decade it had embraced a full range of computers, down to the modest IBM PC. Many thought that this was folly on the part of IBM, but in the short run the IBM PC turned out to be quite successful. However, as PC clones began to appear and IBM erred in its approach to software for the PC (see Chapter 11), IBM began a downward slide. As it moved into the 1990s, the company was losing billions each year and was in need of a new approach and new leadership. This was achieved in the person of Louis Gerstner [5.9], who broadened the view of the company yet one more time. IBM became a true systems integrator, bringing a full range of hardware and software solutions to its customers instead of just IBM and software. As the twenty-first century was beginning, IBM had largely recovered. One can attribute its continuously broadened view to much of its success over the years.

A broadening of IBM's perspective might be construed to be the points at which they fully embraced the transition from high-end mainframe hardware to midrange workstation and personal computer systems, including both hardware and software. Up to that point their views were arguably quite narrow, and they paid a significant price for those restricted views. For example, if IBM had fully appreciated the value of software, they would never have made the "sweetheart" deal with Microsoft for IBM's PC software, known as PC DOS. From that "moment of truth" in 1979 [5.10], Microsoft was able to set forth on a journey with a future that might have been captured instead by IBM.

5.5   HALOID'S BOLD STEPS

At one time, Haloid was a company in Rochester, New York, that lived in the shadow of Eastman Kodak. It produced photographic paper and got its name from silver halide, a chemical used in photographic processing. Its greatest success appeared to be a very high quality photo paper called Haloid Record. But Joe Wilson, who became president at age 37, was in search of new, potentially breakthrough products. After much searching, Wilson gained the rights to patents on a new copying process, called *electrophotography*, controlled by the Battelle Memorial Institute in Columbus, Ohio, through its original inventor, Chester Carlson. Wilson bet the company on this process and the patents, to which he gained all rights by 1956.

One can argue from these actions that Wilson broadened his view of his business as well as his intended markets. And this wide perspective led him to create, with the primary help of Chester Carlson and his patents, along with Sol Linowitz, one of the largest and most successful companies of his time—the Xerox Corporation. The road to this success was not an easy one. There were discussions and problems with the likes of IBM, RCA, and GE, all of which represented threats to the Xerox claims and business. There was a rocky road to international partnering with Rank (U.K.) and Fuji (Japan). But on the back of its primary machine, the 914 copier, company revenues grew from $33 million in 1958 to $176 million in 1963 [5.11]. By 1966 the market value of the company's stock had grown to $4.5 billion, twice as large as the value of Chrysler or U.S. Steel. All of this, it can be argued, derived from a broader and more generalized pattern of thinking.

On the other side of the coin, one might argue that Xerox overstepped solid business practices by generalizing too far. For example, it is well known that Xerox entered the real estate business, which turned out to be a disaster for them. For a time, they also seemed to abandon the low end of the copying machine market, opening the door rather widely for several foreign models (e.g., Canon) to take critical positions in that market. Twenty–twenty hindsight suggests that was a serious mistake that was traceable to a narrow view of the overall market, which they had created almost singlehandedly.

5.6 FROM PERT TO GERT TO STOCHASTIC NETWORKS

In the early 1960s, this author was asked to look at PERT (the program review evaluation technique) [5.12] and its relationship to R&D projects and programs. One customer was a federally funded R&D center and the other was an administration within the federal government. After awhile, it occurred to me that PERT had certain limitations that could be overcome by a broadening of one's approach. This led to my generalization of PERT by introducing uncertainty as to which paths might, or might not, be taken. I documented this broader approach in the literature [5.13], which created quite a lot of interest. This all eventually led to GERT (the graphical evaluation review technique) and to the concept as well as practicalities of stochastic networks [5.14]. All of the original PERT methodology then became a "special case" of stochastic network construction, analysis, and use. This rather personal example illustrates how powerful a generalization can be in terms of giving rise to a broad range of new considerations and arenas for application.

5.7 REINVENTING YOURSELF

The fast-moving and generally high-tech world of systems suggests that every person working in that world needs to be thinking about his or her skills, past, present, and future. Well-conceived shifts in the application of current skills as well as the acquisition of new skills are more-or-less always on the table in order to be relevant in, and responsive to, the twists and turns in the marketplace. Clever shifts in the application of current skills are part of such deliberations, as the following example illustrates.

Consider the case of John, who as an engineer had worked for years in the field of modeling and simulation (M&S). He had built several simulations from scratch and also became expert at applying many COTS simulations, such as GPSS, SLAM, Simfactory, and others [5.15]. John was a valued employee, but he also noticed that in recent years, the demand for his various skills appeared to be lessening. Then came the year 1993 and the appearance of a blockbuster book that introduced the notion of *business process reengineering* (BPR) [5.16]. Within a relatively short period of time, both government and industry seemed to embrace this notion, starting new programs with new funding. BPR was revolutionary, arguing what turned out to be a simple idea: If you don't like the results you're getting, change the process that led to those results. This, in turn, meant that we needed to define processes in considerable detail and to be able to model these processes so that we could accurately predict the change in results as processes were changed. John was insightful enough to see how his modeling and simulation skills gave him a strong position in this new field of BPR. He was clearly able (1) to model a wide variety of processes and (2) to use simulation software to determine process behavior and results in very accurate and quantitative terms. As he helped his company see the connection between M&S and BPR, he also reinvented himself as part of the leading edge of the company's foray into BPR consulting.

How does all of this apply to the perspective of broadening and generalizing? In this case, John was able to see how to broaden into the new field of BPR using his rather strong skills originally developed in other fields. By "reinventing" himself in this particular way, he set the stage for years of further progress in his career. Along the way, John was also able to develop new skills that were related to better understanding of a wide range of business processes. All of this was facilitated by his ability to broaden and generalize. This example illustrates the value of this approach at the level of the individual and demonstrates how powerful it can be in terms of one's career choices.

5.8 SUMMARY: A MEETING

The notion of broadening and generalizing as a particular way to think outside the box is further illustrated by placing it in the context of a hypothetical meeting in a high-tech company. This same approach is taken as well for each of the remaining eight perspectives with regard to ways of thinking outside the box.

This is an account of two meetings from the background of this author, the first of which happened, the second of which did not. It seems that a company was bidding on a contract with the federal government to build a major communication system. At a key architectural design meeting, they decided rather quickly to pursue a frequency-division multiplex (FDM) approach, not seriously considering other alternatives. They won the contract, and all was well for about two years, when a phase of work was completed. The customer proceeded with its original plan to open the next phase to a new set of competitive bids. As it turned out, a new competitor appeared and proposed a new architecture for the system using a time-division multiplex (TDM) approach. Since the world was "turning digital," the government bought that approach and the original company was out in the cold.

The presentation above is an overview of what actually happened. Now we ask ourselves, is there a scenario that could have led to success by employing this first way to think outside the box, that is, by broadening and generalizing?

In this new scenario we start at the point where the government decides to rebid the next phase of the contract. At the crucial meeting to consider the fundamental approach that will be the centerpiece of the proposal, the company decides to broaden that approach. It recognizes the fact that digital technology is becoming more important and could be a better solution. What to do? Instead of proceeding steadfastly with FDM, the company offers a broader view. This is a formal option for the government to go to TDM, suggesting that if the government selects that approach, the company (1) knows how to do that, and (2) is uniquely suited to figure out how to use maximally the results of the first two years of work on the program (instead of just throwing them away). By broadening its approach, the company was thinking outside the box, thereby creating an attractive alternative for the customer. Since offering a second option constitutes a legitimate proposal, the company was, in effect, covering both bases. Under this fictitious scenario, the company might have won the next phase, proceeding with the customer into the world of digital communications. It had to be more work to write the proposal, but it was worth it. It made the difference between success and failure. Unfortunately, at that crucial meeting, the success scenario was not given a chance to work.

REFERENCES

5.1   Eisner, H. (2000). *Essentials of Project and Systems Engineering Management*, 2nd ed. New York: Wiley.

5.2   Eisner, H. (2003) Eisner's Architecting Method (EAM): Prescriptive Process and Products, Tutorial H02, presented at the 13th Annual International Symposium, INCOSE Arlington, VA, June 29–July 3.

5.3   Sage, A., and J. Armstrong, Jr. (2000). *Introduction to Systems Engineering*. New York: Wiley.

5.4   International Council on Systems Engineering, www.incose.org.

5.5   Eisner, H., J. Marciniak, and R. McMillan (1991). Computer-Aided System of Systems (S2) Engineering, presented at the IEEE International Conference on Systems, Man, and Cybernetics, Charlottesville, VA, October 13–16.

5.6   Keating, C., R. Rogers, R. Unal, D. Dryer, A. Sousa-Posa, R. Safford, W. Peterson, and G. Rabadi (2003). System of Systems Engineering, American Society of Engineering Management, *EMJ Engineering Management Journal*, Vol. 15, September.

5.7   Maier, M. (1998). Architecting Principles for Systems of Systems, *Systems Engineering*, Vol. 1, pp. 267–284.

5.8 Sage, A., and C. Cuppan (2001). On the Systems Engineering and Management of Systems of Systems and Federations of Systems, *Information, Knowledge and Systems Management*, Vol. 2, No. 4, pp. 325–345.

5.9 Gerstner, Louis, Jr. (2002). *Who Says Elephants Can't Dance?* New York: HarperBusiness.

5.10 Cringely, R. (1992). *Accidental Empires*. New York: HarperCollins.

5.11 Linowitz, S. (1985). *The Making of a Public Man: A Memoir.* Boston: Little, Brown.

5.12 Malcolm, D., J. Roseboom, C. Clark, and W. Fazar (1959). Application of a Technique for Research and Development Program Evaluation, *Operations Research*, Vol. 7, No. 7, pp. 646–669.

5.13 Eisner, H. (1962). A Generalized Network Approach to the Planning and Scheduling of a Research Project, *Operations Research*, Vol. 10, No. 1.

5.14 Whitehouse, G. (1973). *Systems Analysis and Design Using Network Techniques.* Englewood Cliffs, NJ: Prentice-Hall.

5.15 Eisner, H. (1988). *Computer-Aided Systems Engineering.* Englewood Cliffs, NJ: Prentice Hall.

5.16 Hammer, M., and J. Champy (1993). *Reengineering the Corporation.* New York: HarperBusiness.

Chapter 6

Perspective 2: Crossover

Perspective 2 involves creating leverage. One way to do this is to build a system once and use it many times. Leverage through multiple sales and use is conventional wisdom in some domains, but it is not fully appreciated in many others. All companies that make consumer products accept multiple purchase and use as an axiom. Anywhere from selling automobiles to cans of soup, the object is to sell as many as possible. However, in the world of systems, the situation becomes less clear, especially under the procurement regulations and practices of the federal, state, and local governments. In these domains, companies usually bid on contracts to build a system. They also tend to start each such development with a "clean sheet of paper," attempting to construct a system that will satisfy every single customer requirement. If several "copies" of a system are needed, the price for each copy is usually part of the written contract. The name I've given to this particular way to think outside the box is *crossover*. Typically, crossover creates leverage by making a system (or part of a system) once and utilizing it many times.

Here are a few examples of how crossover leverage may be created:

1. Build an inventory control system for the Navy, and then provide it to the Army and the Air Force.
2. Build a legislative tracking system for the federal government, and then sell it to the state and local governments.
3. Build a human resources utilization system for the plastics industry and then sell it to the transportation and heavy machinery industries.

Managing Complex Systems: Thinking Outside the Box, By Howard Eisner
Copyright © 2005 John Wiley & Sons, Inc.

These are somewhat idealized, but they serve to make the point.

Another aspect of this approach is to try not to "reinvent the wheel." Once a good solution is found, stick with it. A policy of continuous improvement will help make the wheel better, in evolutionary but significant steps. Creating an entirely new wheel each time is usually (but not always) wasteful rather than highly productive.

6.1 PEOPLESOFT

Peoplesoft is a company that does a lot of business with the federal government. It began by designing and building fairly straightforward information systems to the specifications of a government customer. Such "plain vanilla" systems were generally non-real-time applications in such domains as human resources and personnel management. As they built systems in this way for various agencies, they realized that the requirements did not differ greatly from agency to agency. Further, by developing a deep understanding of the needs described by these agencies, they could (1) focus on a core or common set of needs, and (2) begin to market their systems to virtually all of the executive branch agencies. As a consequence of this "make once, sell many times" philosophy, they began to achieve success that surpassed the approach taken by companies used to the idea that every system is custom made for the unique requirements of each customer.

In its idealized form, Peoplesoft could build a very good human resources tracking system for one department, and then sell it, in perhaps a slightly modified version, to another 25 departments in the federal government. This resulted in lower costs to the customer, less money spent on custom proposal writing, and a more than rudimentary application of the principle "make once and sell many times." So here is a latecomer to the federal market, taking an out-of-the-box approach. By so doing, they outperformed companies with 20 to 30 years of experience in building systems for the government. For many months of 2004, Oracle tried to acquire Peoplesoft [6.1], making several offers of approximately $7 billion to $9 billion. The board of directors of Peoplesoft turned down all these offers. Finally, by December 2004, the board agreed to an offer of about $10.3 billion. Judging from this experience as well as other business data, it appears that there is considerable merit to a successful employment of the crossover approach. With its better idea, Peoplesoft did not try to buck or fight the way in which the government acquired systems. Instead, they had an out-of-the-box approach which they could adapt to the vagaries of how the federal government does business.

6.2 SOFTWARE REUSE

Another example of crossover is embedded in the various aspects of software reuse in the building of systems. As we have accepted both the idea and the practice of reusing hardware designs that work, software reuse is becoming increasingly important as part of the way we build systems. Apparently, many software designers and programmers would just as soon start with the proverbial "clean sheet of paper." Such an approach tends to look a whole lot like reinventing the wheel. Moving forward with software reuse involves planning and organization, but the potential reward of delivering software systems at greatly reduced costs is extremely attractive.

An example of the results of software reuse can be found in Poulin's examination of this field [6.2]. He articulates the expected benefits of reuse, including:

1. Enhanced productivity
2. A reduction in risk
3. Reduced costs
4. Better interoperability
5. An increase in quality
6. Lower cycle times

In addition, based on his review of earlier studies, he declares that software reuse will require only 20% of the effort of a new development. This is clearly a significant saving if it can be achieved consistently; and it is a concrete as well as a quantitative expression of what is meant by *creating leverage*.

Another exploration of software reuse [6.3] identifies specific phases in which savings may be realized, including:

1. The requirements and specification phase
2. The design phase
3. The coding phase
4. The integration and test phase
5. The maintenance phase

In addition, the above author recognized that savings may be possible through a rather large number of reusable artifacts, listing some 18 such artifacts that relate, in one way or another, to the process of software development. Also, specific attention is paid to reuse libraries, whose purpose is to facilitate the matter of software reuse.

In 2004, a doctoral dissertation addressed a variety of software reuse matters [6.4], with the following selected results from a questionnaire:

1. More than 90% agreed that software reuse was accepted within their organization and that they preferred to reuse rather than not.
2. Almost 90% agreed that they were able to reuse software and that they had a positive experience with software reuse.
3. Almost 90% agreed that reuse pays for itself.
4. 98% agreed that there was a benefit from software reuse

5. The top four reasons given for the reuse of software were:

   a. Productivity increases

   b. Reduction of costs

   c. Increases in quality

   d. Reduction of cycle times

Recognizing these types of benefits from software reuse, the government has been supporting reuse, although in many ways it has a new built-in problem to contend with. If two firms bid different amounts of software reuse on a proposed system, significant differences in the prices proposed are likely to result. There have been several cases where the government selected the winner to be the firm that promised the greatest amount of reuse. When this amount of reuse did not materialize, there were few, if any, penalties. A better way of dealing with these types of situations is still considered to be a work in progress.

6.3 DEVELOPER OFF-THE-SHELF SYSTEMS

In 1995 this author introduced the notion of revising the software acquisition process by using developer off-the-shelf systems (DOTSS) [6.5]. A progress report on this concept was provided two years later [6.6]. The DOTSS notion is related directly to the practice of software reuse, with the major change being that the *entire software system* is being reused instead of selected portions thereof. For this to be workable, the domain of application is limited to "plain vanilla" non-real-time information systems such as human resource tracking, inventory control, financial management, and similar types of systems. The reason is that most of these systems have identical functionality. Further, it is this type of approach that has led to much of the success of PeopleSoft, as described earlier in this chapter and in the literature.

The significance of the DOTSS approach lies in the crossover and leverage that it provides. After building and installing a state-of-the-art software system for one agency, one is in a position to use this proven system to crossover to another agency (in fact, several other agencies). We can obtain an overall estimate of the leverage thus accrued by noting that through this large-scale reuse, we are able to provide an IT system at a savings of about 80% in *both* cost and time [6.6]. This is measured as one-fifth of the time *and* cost, in relation to constructing a "clean sheet of paper" system, as illustrated in Figure 6.1. This results in a benefit ratio of 25, a 2500% improvement (!). Can you think of a change in process that offered a 25% improvement, much less 100 times that amount?

To gain some perspective regarding what might happen if you develop an out-of-the-box approach to solving a problem, consider some of the real-world responses to the changes implied by DOTSS [6.6]:

1. It's basically a useful notion, and we're already doing it.

2. It's an excellent idea, but it threatens to change 30 years of prior practice and contrary culture.

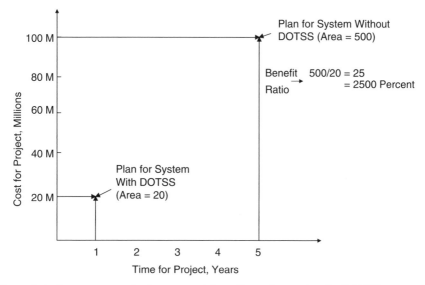

Figure 6.1 Improvements in the cost and time dimensions using the DOTSS approach (illustrative).

3. Each system has unique requirements, which is not an idea that can or should be challenged.
4. DOTSS is a potentially powerful way to save lots of federal dollars, but there is little to no incentive to do so.

Perhaps you now wish to consider some of the downside aspects of embarking on a journey that involves thinking outside the box. There are many nay-sayers, and they need to be considered as the journey proceeds.

6.4 ACCOUNTING FIRM CONSULTATION

Another form of crossover leverage was set in motion when the Big 6 (at that time) accounting forms came up with the idea that they should provide systems consultation services to the firms for which they provided accounting services. The crossover in this case was using an existing customer base to sell additional (products and) services, but of a different type. For the most part, the leverage lay in (1) better competitive positioning that accrued from a positive accounting relationship, and (2) significantly reduced marketing expenses in order to capture a "new" client. Many consulting firms charged high prices, making them even more profitable in relation to competitors that did not have this cozy relationship.

Although this crossover relationship persisted for several decades, abuses in the implementation showed themselves, and several accounting firms gave up the practice of seeking nonaccounting business. The practice, more or less, always had a built-in type of conflict of interest. When the perceived conflicts started to become real and undeniable conflicts, some type of reform had to result.

6.5 BUILDING AND MANAGING NEW SYSTEMS

We now come to the matter of what might be done by a systems engineering team that is engaged in some aspect of building and managing a new system for a specific customer. It also applies significantly to the proposal phase since increased leverage normally results in less cost, which, in turn, is likely to affect the price that is proposed by an offeror.

The straightforward answer is that every firm needs to try to position itself to obtain leverage in every aspect of how they do business. Leverage can translate into a reduced cost to the client, which increases the chances of capturing a particular contract. Leverage can also be represented by a better product or service at the same price. So the next question is: How is the leverage to be obtained as well as sustained? Some areas in which crossover leverage may be found include the following:

1. Software development expertise
2. Use of modeling and simulation software from earlier contracts
3. More competent engineers and scientists
4. Training and education programs that lead to area 3 above
5. Higher pay and perquisites that lead to area 3 above
6. An active knowledge management and utilization program
7. Superior methods in systems engineering
8. Superior methods in program and project management
9. Greater engineering maturity
10. Better team building and teamwork in all operations

6.5.1 Software Development Expertise

It has been demonstrated that software development is an area in which there can be a considerable amount of leverage. The leverage inherent in software reuse was discussed earlier in the chapter. This related to the software "product" and its potential use. This discussion pertains more to the software development process that is employed by a software development team.

The leverage that can accrue from a software development team results from extremely high productivity. This means that a person-month of effort put forth by such a team is able to produce more and better software than an average or poor team. High-performance software teams can therefore outperform mediocre teams by significant amounts. In this regard, thinking outside the box requires (1) a definitive decision to invest in high-performance teams (HPTs), and (2) knowing how to build a HPT.

There is also considerable evidence that the investments made with respect to achieving appropriate software capability maturity (CMM) levels have been good ones. Capability levels have increased, and many companies, especially those building software systems for the federal government, have experienced significant returns on their investments. The 18 key process areas that were part of the original software CMM follow [6.7]:

Level 2

Requirements management
Software project planning
Software project tracking and oversight
Software subcontract management
Software quality assurance
Software configuration management

Level 3

Organizational process focus
Organizational process definition
Training program
Integrated software management
Software product engineering
Intergroup coordination
Peer reviews

Level 4

Process measurement and analysis
Quality management

Level 5

Defect prevention
Technology innovation
Process change management

It is easy to see how improvements in these key areas will lead to better ways to build software, itself a crucial part of constructing new complex systems.

6.5.2   Use of Modeling and Simulation Software

Companies that have built modeling and simulation (M&S) software for one customer should be well positioned to reuse that software for another customer. This reuse creates leverage and has the potential for beating the competition on future bids and efforts.

As an example, companies in the business of planning, designing, and building metro-area transit systems (subways) for various cities have, at times, built a digital simulation of the system they are analyzing and designing. This allows them to "run the railroad" on a computer as they try different configurations involving such design features as location of stations, size of cars, train headways, acceleration–deceleration profiles, and others. The simulation software becomes a very valuable design and analysis tool which the company has built for this purpose. Having done so, the software can then be used again for the next contract with a city that wishes to have such a system. Indeed, companies without this type of capability might well have a difficult time being competitive in this domain.

The concept described above is valid in many fields that basically require the use of modeling and simulation to develop answers with respect to design and performance. Following are some of the fields in which M&S software is being used extensively:

- Factory processes
- War gaming
- Movie animation
- General systems operations
- Local area network behavior
- Railroad operations
- General network analysis
- Assembly line operation

6.5.3   More Competent Engineers and Scientists

The area of competent engineers and scientists refers, of course, to the "people" side of the equation. There is both crossover and leverage in the selection of the right people to design and build the system at hand. Perhaps the simplest example of crossover can be seen by referring back to the beginning of this chapter:

- Build a system for customer X and then sell it to customer Y.

Along with the notion above is a parallel idea:

- Use several of the key people who built the system for customer X to build the required system for customer Y.

Having had the experience with the first customer, crossing over to the second is very likely to be extremely cost-effective, with numerous opportunities to be more efficient. As alluded to earlier, this may be multiplied further if the system consists of software only.

Some enterprises make a fundamental mistake in this area. They believe that competent engineers and scientists do not like to do the same job twice and then apply it to the crossover scenario. To that we would respond by saying that no two jobs are precisely the same, and there is considerable merit to having hands-on experience with the next job coming down the road. Further, having this type of competence on the next job places the company in the position to make more money (fixed-price basis) and thereby share some of that benefit with the same engineers and scientists. Another positive choice is to use the excess dollars to build a better product for the customer. These are excellent choices for a company to be able to select.

6.5.4 Training and Education Programs

In 1990, Peter Senge argued that all modern enterprises should strive to be *learning organizations* [6.8], and by so doing they will "continually enhance their capacity to realize their highest aspirations." A very conventional as well as correct response to this suggestion is that today's enterprises need to be sure that they have active and well-focused training and education programs for their employees. Training tends to address quite specific skills that need to be mastered, whereas education deals more with academic programs that lead to a degree. Both are recommended to meet the demands made on both business and government in today's world. A less conventional response to Senge's admonition is to assure that the enterprise learns the five disciplines that he prescribes, as cited in Chapter 3.

To Senge, the enterprise needs to focus on the five disciplines in order to become a learning organization. Indeed, the last of the above is the fifth discipline (the title of Senge's book) as well as one of the nine ways of thinking outside the box presented in this book (Chapter 13). Each of the five disciplines can also be thought of in terms of three distinct levels: practices, principles, and essences.

Two other points made by Senge are worth special attention here. The first is that it is possible to have small changes produce large and significant results. However, he points out that the areas in which the highest leverage may be obtained can often be the least obvious. The previously presented notion of DOTSS presented here is one example in which a great deal of leverage is obtained by a small change in one's thinking about the value of what an enterprise has already achieved. Another point made by Senge is that there is a core learning dilemma, the "delusion of learning from experience." This requires considerable more explanation, one aspect of which is that we often do not get a chance to make appropriate observations about decisions we have made in relation to their consequences as played out in the real world. We must go beyond our direct experiences into the five disciplines that the author suggests.

6.5.5 Higher Pay and Perquisites

All companies have access to job-related data that show what salaries are for various positions. Often, these data take the form shown in Figure 6.2. For a given job classification or type, the data tend to display salary as a primary function of years of experience, with percentiles as a parameter. This information shows a spread of salaries for a given number of years. Looking at the 20-year point, we see

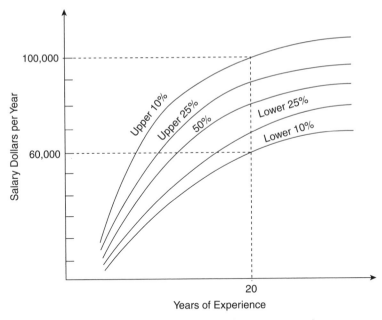

Figure 6.2 Typical salary curves for years of experience and percentages.

an upper 10% salary of $100,000 and a lower 10% salary of $60,000, with several points in between. The question for all companies then becomes: How do we use these curves to establish the pay levels for our people? Despite the apparent conjecture that not all people are above average (the 50% curve), companies should strive to populate their workforce at the higher levels. By paying more (on average) than your competitors, you are likely to get better people. The same is true in the area of perks.

The next question one might ask is: Can we afford to do this? The argument that supports this high-end approach is that you believe the higher pay brings better people, which in turn brings higher productivity and leverage that *more than* justifies the additional costs. If you are in a business where this position cannot be supported, don't go there. However, leverage is often created by people (vs. process, for example), and it is the best people who are capable of creating the leverage. Of course, they can't do it in a vacuum. Once they have been hired, the enterprise needs to be listening to and accepting what they have to offer. What they tend to offer is an ability to think, as well as produce, outside the box.

6.5.6   Knowledge Management and Utilization

Enterprises today are, more and more, appreciating the fact that if they can better understand what knowledge they have, as well as do not have, they are more likely to be competitive, other factors being equal. This has legitimized the field of knowledge management and the utilization of this knowledge.

Knowledge management can be enhanced if viewed through the prism of crossover and leverage. Indeed, part of knowledge management is to apply efficiently the knowledge that an enterprise has from one domain to another. Here are just a few examples:

1. We know how to build human resource tracking software and should be able to apply that knowledge to document status tracking.
2. We know how to build a model of the software development process and should be able to apply that knowledge to the systems engineering process.
3. We know how to simulate the behavior of an assembly line and therefore should be able to apply that knowledge to defining operating profiles for a ground-based satellite tracking station.

Knowledge crossover, properly applied, can create leverage and high productivity. Figuring out how to capture and then apply the knowledge is a problem that thousands of companies seem to be working on today. They would not have such programs if the payoff was perceived to be marginal or nil.

6.5.7 Superior Methods in Systems Engineering

If we are able to carry out the tasks of systems engineering more effectively and efficiently than our competitors, we will be creating a leveraged advantage that is likely to lead to success. This perspective should be even stronger in the future as systems become larger and more complex. The tasks of systems engineering have been defined by many authors and also with respect to the construction of a systems engineering capability maturity model (SE-CMM) [6.9]. In the main, the key process areas for systems engineering have been subsumed into the integrated CMM (CMMI), which is discussed in somewhat greater detail later in this chapter.

Additional insight into systems engineering and its potential value was provided by Kludze in his exploration of systems engineering at the National Aeronautics and Space Administration (NASA) [6.10]. Here are some of the results of his examination, which are listed by NASA employees as well as members of the International Council on Systems Engineering (INCOSE):

1. Ranked cost–benefits from the use of systems engineering as good or excellent:
   a. NASA = 76%
   b. INCOSE = 84%
2. Systems engineering helps in developing cost-effective systems (agree or strongly disagree):
   a. NASA = 83%
   b. INCOSE = 91%

3. Overall impact of systems engineering (some positive impact):
   a. NASA = 95%
   b. INCOSE = 94%

Most other results were similar to the above with respect to such factors as effects on schedule, value in terms of meeting system objectives, and use on future projects. This investigation clearly supports the notion that systems engineering is a critical aspect of building and managing successful large-scale complex systems.

6.5.8 Superior Project Management Methods

One might say that managing a project is one of the most important and fundamental aspects of management in general. A project is where the "rubber meets the road": A group of folks need to be able to provide a service or product to a particular customer within budget and time constraints, and satisfying specific performance requirements. Projects vary a great deal, especially with respect to size. A three-person project is usually quite different in many respects from a 300-person project, even though they both address the basic functions of planning, organizing, directing, and monitoring [6.9].

Since project management is a basic unit of management and is ubiquitous in most organizations, small increases in efficiency at the project management level will propagate throughout, creating significant benefits. This is often not appreciated, and projects are set in motion without the necessary training and education. Some project managers grow into the job, often using sheer intuition as a method and perseverance as a guideline. This approach is usually not good enough. The learning organization figures out that (1) project management is an important area that needs to be addressed, and (2) project management should always be on the list for continuous improvement.

6.5.9 Greater Engineering Maturity

The area of greater engineering maturity relates to the capabilities of an enterprise with respect to its overall processes and products, from the point of view of building and managing complex systems. To the extent that this capability is current, active, and superior, it will create leverage in the marketplace. If it is spotty and largely incoherent, it is likely to be surpassed by other organizations. Achieving the desired capability can also be thought of as being dependent on two well-known ideas:

1. Becoming a learning organization
2. Maintaining a consistent practice of continuous improvement

The capability maturity model (CMM) notions discussed earlier in this chapter have been generalized to the point where they apply to at least three distinct domains:

1. Systems engineering
2. Software engineering
3. Integrated product and process development

Such a model is the integrated capability maturity model (CMMI) [6.11, 6.12]. This model defines a set of process areas (PAs) for the various maturity levels shown below (*, engineering PAs; **, project management PAs; ***, support PAs; ****, process management PAs).

Maturity Level 2 (Managed)

Requirements management (*)
Project planning (**)
Project monitoring and control (**)
Supplier agreement management (**)
Measurement and analysis (***)
Process and product quality assurance (***)
Configuration management (***)

Maturity Level 3 (Defined)

Requirements development (*)
Technical solution (*)
Product integration (*)
Verification (*)
Validation (*)
Organizational process focus (****)
Organizational process definition (****)
Organizational training (****)
Integrated project management (**)
Risk management (**)
Integrated teaming (**)
Integrated supplier management (**)
Decision analysis and resolution (***)
Organizational environment for integration (***)

Maturity Level 4 (Quantitatively Managed)

Organizational process performance (****)
Quantitative project management (**)

Maturity Level 5 (Optimizing)

Organizational innovation and deployment (****)
Causal analysis and resolution (***)

This rather long list of process areas is certainly a comprehensive structure for dealing with the important capabilities that need to be present to achieve success in today's complex world.

6.5.10   Better Team Building and Teamwork

Our last item on this list of key leverage areas in building and managing new systems is that of constructing a team to develop these systems. As we know, a team is more than just a group of people who have been assigned to a new project. Team building and operation can clearly be a *force multiplier* such that the productivity of a team is greater than the simple summation of individual productivity measures. Members of a true team are excited about what they do, especially how they do it together.

Senge also talks about *team learning*. His perspective regarding the learning organization and the five key disciplines has not overlooked the important role of the team. Without the team, much of what Senge envisions just simply does not happen.

The reader is also referred to other sources [6.9] as well as to Chapter 14 for additional information regarding the operation and value of teams. Teams provide critical leverage that is part of this particular perspective for thinking outside the box.

6.6   SUMMARY: A MEETING

We now come to a hypothetical meeting held at the ATC (Advanced Technology Corporation) as they were trying to complete their strategy on a bid to the federal government. They had put together a proposal for a financial management system (FMS) needed by an important government agency, and the key players were meeting to do a final review of their overall approach. In attendance were:

George: the company's senior vice-president
John: the company's vice president in charge of this line of business
Cindy: the company's chief technology officer
Blake: the proposal manager for this project
Ruth: the chief software development engineer

George: I called this meeting so that we could take a last look at our approach to this important bid for the company. As I understand it, we are putting the final touches on our proposal for this FMS, and we have a cost estimate for the project of approximately $62 million. Is that where we are?

Blake: That's about it, along with a 36-month time frame.

George: Do we all think this is a winning strategy?

John: I'm not sure that it is, but we are bidding the explicit reuse of software from our recently completed project, the financial accounting and management system, for another agency.

George: And how much reuse do we expect to get?

Ruth: The precise number is about 38%.

Cindy: And that translates into a significant dollar savings.

George: And do we all think that this is our best position? That this will lead to a win?

John: Speaking for myself, I don't feel confident that this bid will win the contract. The ABC company, always a strong competitor, may be able to beat us on this one.

Ruth: I agree. I think we have a very strong proposal, but it still may not be the winning one.

George: Well, this is our final run at this. Does anyone have a strategy in mind that will, for example, definitely place us in the number one position?

Cindy: I've got an idea that I'd like to put on the table that might do that.

John: OK, sure. Let's hear it.

Cindy: We've bid what amounts to a software development job, with about 38% software reuse. That moves us down the field, but our friendly competitors at ABC might well take a similar approach but bring the software reuse to 50%. In that case, everything else being equal, they might just walk off with this contract on the basis of cost.

George: Do you have a way of guarding against that?

Cindy: I think I might. Why not come in with a DOTSS approach—not as our primary bid but as a second option. When I ran through the numbers, that approach will deliver a working FMS for about $11 million.

John: But we've discussed this before. Our FMS system meets about 90% of the requirements. It doesn't get us all the way to 100%.

Cindy: I know that, we all know that. But here's the value proposition that we can put to our customer: Our primary bid is for $43 million, and it will meet all your requirements and include some 38% reuse. At the same time, we are offering a second option, based on the DOTSS idea, in which we will deliver a working FMS at a cost of $11 million that will meet 90% of your requirements. Further, that working system will be our just completed FAM system for a different agency. We can also show that the missing 10% of the requirements is not really consequential. Further, this approach can be accomplished in 12 months instead of 36 months. Which alternative would you like?

Blake: Is that a legal bid?

Cindy: Under a best-value acquisition approach and the fact that alternative or optional bids are acceptable, as I understand it, all of this is legal.

John: And can we substantiate this approach in black and white so that both we and the customer find it workable?

Ruth: We've done our homework on this, and I think the answer to that question is "yes."

George: I like it. It's "out of the box" but it stands a chance of being successful.

John: If I were the customer, I'd go for it.

George: Then let's do it. Do we have any negative votes?
[None are expressed.]

George: Many thanks to you all. I think we're in the right place on this. Let's do it!

So the company went down the road with an out-of-the-box approach as a second optional bid. Did they win the contract? We'll leave that question for you, the reader, to answer.

REFERENCES

6.1 Liedtke, M. (2004). Acquisitions Essential, Oracle CEO Asserts, *Washington Post*, July 1.

6.2 Poulin, J. (1997). *Measuring Software Reuse*. Reading, MA: Addison-Wesley, Longman.

6.3 Leach, R. (1997). *Software Reuse*. New York: McGraw-Hill.

6.4 Gromadzki, R. (2004). Extent and Issues of Software Reuse, Ph.D. dissertation, The George Washington University, May 16.

6.5 Eisner, H. (1995). Reengineering the Software Acquisition Process Using Developer-Off-the-Shelf Systems (DOTSS), presented at the IEEE International Conference on Systems, Man, and Cybernetics, Vancouver, British Columbia, Canada, October 22–25.

6.6 Eisner, H. (1997). The DOTSS (Developer-Off-the-Shelf Systems) Approach: A Progress Report, presented at the Portland International Conference on Management of Engineering and Technology (PICMET '97), Portland, OR, July 27–31.

6.7 Paulk, M., B. Curtis, and M. B. Crissis (1991). *Capability Maturity Model for Software*, CMU-SEI-91-TR-24. Pittsburgh, PA: Software Engineering Institute.

6.8 Senge, P. (1990). *The Fifth Discipline*. New York: Doubleday Currency.

6.9 Eisner, H. (2002). *Essentials of Project and Systems Engineering Management*. New York: Wiley.

6.10 Kludze, A. (2003). Engineering of Complex Systems: The Impact of Systems Engineering at NASA, Ph.D. dissertation, The George Washington University, Engineering Management and Systems Engineering Department, August 31.

6.11 Ahern, D., A. Clouse, and R. Turner (2001). *CMMI Distilled*. Reading, MA: Addision-Wesley.

6.12 Software Engineering Institute (2003). Integrated Capability Maturity Model (CMMI), version 1.1, wallchart. Pittsburgh, PA: SEI, Carnegie Mellon University.

Chapter 7

Perspective 3:
Question Conventional Wisdom

It can be argued that essentially all of our creators and inventors of something new, from a better idea to its physical manifestation, have been able to question conventional wisdom. Depending on the field of endeavor, it normally takes a considerable amount of courage to question conventional wisdom. People are often not interested in hearing about someone else's new theory, or even a better way to do business. Much is invested in the old ways, for reasons that are easy to imagine. But this third perspective in regard to thinking outside the box focuses specifically on the matter of questioning conventional wisdom.

Following is a short list of people, from inventors or scientists to business people, who were sufficiently able to question conventional wisdom that their stories have often been told and are well known:

Inventors/Scientists	Businesspeople
R. Feynman	L. Iacocca
T. A. Edison	T. Watson, Jr.
J. Salk	W. Hewlett and D. Packard
R. Goddard	A. Grove
I. Sikorsky	R. Perot
C. Shannon	N. Augustine

Managing Complex Systems: Thinking Outside the Box, By Howard Eisner
Copyright © 2005 John Wiley & Sons, Inc.

Looking at the list of inventors and scientists, we come first to Richard Feynman, a physicist and Nobel laureate who also wrote several trade books in which he tried to explain a very difficult subject to laypeople. Thomas Alva Edison, the Wizard of Menlo Park, New Jersey, was one of our most eminent inventors, with 1093 patents that had his name on them. In addition to the light bulb, he is usually credited with inventing such devices as the universal stock ticker and the phonograph. Jonas Salk is well known for creating a successful vaccine for polio. This type of creativity can be immediately traceable to the saving of millions of lives. Robert Goddard, for whom NASA's Goddard Space Flight Center is named, is well known for his development of the modern field of rocketry. In his case, it *was* rocket science. Igor Sikorsky was a leading creator of helicopters and associated technologies. Finally, Claude Shannon is largely credited with creating the field of information theory. This, in turn, has served as a foundation for many of the coding principles in various aspects of communications engineering.

On the business side, Lee Iacocca came into national prominence with his marketing and business leadership at the Ford Motor Company, starting with the success of the Mustang. He later helped to lead the Chrysler Corporation during difficult times. Tom Watson, Jr., mentioned in Chapter 5, was instrumental in bringing IBM into the large mainframe computer world, from its earlier days in tabulating machines and typewriters. Bill Hewlett and David Packard created the Hewlett-Packard Company, working from a garage in Palo Alto, California. Their special way of dealing with people and products led to many years of consistent success in the high-technology field. Andy Grove took the helm at the Intel Corporation and made it the most successful chip manufacturer in the world. Ross Perot was a business leader who had a distinctive style that helped him create Electronic Data Systems (EDS) (which he sold to General Motors for a handsome price) as well as its successor, Perot Systems, Inc. He has used these successes to move into the political arena. Finally, Norman Augustine has held leadership positions in both Martin Marietta and Lockheed Martin. He is well known for his understanding of defense matters as well as for his contributions to worthy causes (see also Chapter 4).

The short discussion above has highlighted people who were able to question various aspects of conventional wisdom, leading to extraordinary results. They give us an idea of what might be possible by having the courage and the intellect to move forward in areas where others might not have been able to do so.

7.1 LARGE AND COMPLEX GOVERNMENT SYSTEMS

The executive branch of our government has often taken the lead in developing large, complex systems, with the Department of Defense (DoD) showing the way. This leading-edge position has been largely born of necessity; we have needed systems that give us clear superiority over our enemies. In moving down this road, we have had to create as well as challenge conventional wisdom many times over. The following sections provide examples drawn largely from this domain.

7.1.1 Can You Hit a Bullet with a Bullet?

Since around 1981 we have been examining the question: Is it possible to hit a bullet with a bullet? The question was not proposed idly or as an academic exercise. It was front and center in the Strategic Defense Initiative program and continued later as the National Missile Defense program. The key issues have massive defense as well as political dimensions and lots of people on both sides of these issues. One might say that the conventional wisdom has been that it can't be done, at least not now, with a sufficiently high success rate. The questioners of that wisdom claim that it's just a matter of time before we can. Yet other nay-sayers argue that whether or not it's technically feasible, we shouldn't be spending tax dollars in that way, for more than one single reason.

A system to defend against possible enemy missile attacks is quite a complex system. It would certainly appear to be our most complex air defense system, consisting of at least the following top-level functional elements:

1. Air surveillance
2. Target detection
3. Tracking
4. Identification
5. Assignment of missiles
6. Launch
7. Guidance
8. Command and control
9. Missile kill
10. Damage assessment

This knotty example illustrates, in today's complicated world, why it can be so difficult to question conventional wisdom. The issues are complex and the time lines can be long, as we make progress rather slowly. So what *is* the answer to the question of whether we can hit a bullet with a bullet. At the moment, the answer would appear to be yes under some circumstances, and no under other circumstances.

7.1.2 Commercial Off-the-Shelf Systems

There was a time when the DoD kept away from COTS systems, under the general perception that they could not meet stringent military requirements. By questioning that particular form of conventional wisdom, over time, the DoD has changed its mind. Indeed, this change is so complete that the use of COTS is explicitly encouraged [7.1]. This is one way in which new conventional wisdom is formed. Old conventional wisdom is rejected and the articulation of that rejection becomes a form of the new conventional wisdom. As so the wheel turns as we change our minds about what works and what does not work.

7.1.3 The Integration of Systems

The conventional wisdom would appear to be that more integration is always better than less integration. This piece of conventional wisdom also applies to what to do with stovepipe systems. To question this conventional wisdom (see also Chapter 1) might be to say there is a degree of integration that might be called optimal, and this degree varies quite a lot, depending on many factors and circumstances. For some situations, 20% integration might be best (most cost-effective), and for others, 70% might be the proper approach. The truth is that this is a very complex issue, and many systems have failed in pursuit of conventional wisdom. In this context, failure is interpreted as not meeting effectiveness goals, or overrunning time or cost constraints. We continue to try to build systems that facilitate integration, but we are far from fully "plug-compatible" hardware and software. We also need to understand that as we climb the ladder toward greater integration, we may also be giving up capabilities, knowingly or otherwise. For example, a failure in one of eight stovepipe systems may put one, and only one, system in a "down" state. With the other seven stovepipes operative, one might say that we are in a degraded mode, with only one-eighth of our capability lost. If all eight stovepipes were fully integrated, depending on how this was achieved, the failure postulated above might cause the *entire system* to go down. Extreme care must be taken to avoid this type of result, thereby creating new dependent failure modes. This could more than defeat the benefits of integration.

7.1.4 Saving Money as We Build Systems

The conventional wisdom in the government as it builds large, complex systems is that saving money is highly desirable. We presume that this perspective is firmly in place and that we are doing as well as we can in this regard. However, the nature of the built-in incentives and processes often conspires against us. Here's about how it works. There is a struggle to get programs funded, often asking for more money than is needed so that the "inevitable" cuts still lead to adequate dollar levels. If these cuts don't occur, no one seems to be going back claiming that we can be okay with less. In short, once a firm budget is established, there is no incentive for the manager to spend less than that if it is possible to do so. Indeed, a manager who wishes to spend less and return monies to the U.S. Treasury would probably have a very short tenure as a manager in the government. The incentives are not there to save money but to spend every nickel that you are authorized to spend, unlike many plans and incentives in the industrial sector.

Government executives and managers gain prestige, of course, the greater the dollar value of the enterprise for which they are responsible. Each budget cycle normally involves a request for more money for each of the next few years. Needing less for next year seems not to be part of the equation. Demonstrating that you can do more with less is usually met with disdain or incredulity. The more money you have to spend is a measure of your importance. Saving money doesn't improve your value; it potentially diminishes it.

In chapter 6, I discussed an initiative in which I participated that could have increased productivity by as much as 2500% (!) under certain circumstances [7.2]. Even after speaking to about 18 high-level managers in the federal government with the sponsorship of one of our leading systems integrators, few showed interest in saving federal dollars. The conventional wisdom may be to save money, but the realities and constraints may lead us to do exactly the opposite.

7.2   CONVENTIONAL WISDOM CHANGES WITH THE TIMES

It is worth noting that there are numerous cases in which yesterday's conventional wisdom has been replaced by today's, which we might expect will be replaced by something else tomorrow. One example is the assembly line. Henry Ford introduced assembly line construction, which superseded one-at-a-time methods. Corporate ownership moved from the very few to the very many over the years. Computers containing hundreds of vacuum tubes were replaced by those using transistors, integrated circuits (ICs), and chips. Companies doing most of their business with the federal government have bounced around quite a bit in terms of their perceived value in the marketplace. From the example above and others, we can articulate some of the factors that force a change in conventional wisdom:

1. Advances in technology
2. Changes in manufacturing methods
3. New business paradigms
4. Economic and industry restructuring
5. Government regulation and deregulation
6. Competition
7. Advances in methods of building large-scale complex systems

Each of these is discussed briefly in the following sections.

7.2.1   The Forces of Technology

Technology improvements represent one of the most powerful forces for change. As such, conventional wisdom is made obsolete to be replaced by what technology is allowing us to do. As builders and managers of complex systems, we must always be prepared to explore in some detail how technology can help us achieve the best results. It is for this reason that the conventional systems engineering management plan contains a section that deals explicitly with technology definition, insertion, and transition [7.3]. An emphasis in technology can also be seen in the defense acquisition management framework [7.1], which has inserted a new technology development phase between concept refinement and system development and demonstration, as shown in Figure 7.1.

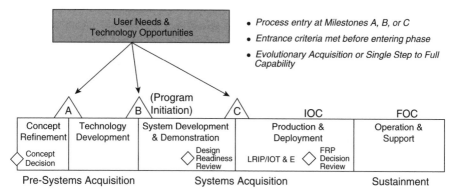

Figure 7.1 Defense acquisition management framework. (From [7.1].)

7.2.2 Manufacturing Methods

At a top level, this point is illustrated by just three words: the industrial revolution. Beyond that, we note that every year we are able to manufacture new electronic chips that are faster and more cost-effective and that, in turn, fuel hundreds of manufacturing processes that are computer controlled and driven. Finally, a critical frontier is how to "manufacture" software. Perhaps this will be the key challenge of the twenty-first century.

7.2.3 New Business Paradigms

There is little question but that new business paradigms challenge conventional wisdom by suggesting that there might be a better way to look at and solve problems. A clear example lies in business process reengineering (BPR), which was articulated in a best-seller in 1993 [7.4]. The message of BPR was and is simple: If you don't like the results a given process is producing, change the process. Other business-oriented paradigms that transcended earlier conventional wisdom include:

1. Drucker's management by objectives (MBO) [7.5]
2. Deming's quality assurance and total quality management (TQM) [7.6]
3. Peters and Waterman's in search of excellence [7.7]
4. Ouchi, Pascale, and Athos's Japanese style of management [7.8, 7.9]
5. Kaplan and Norton's balanced scorecard [7.10]
6. Goleman's emotional intelligence [7.11]
7. Senge's learning organization and systems thinking [7.12]

7.2.4 Economic or Business Restructuring

Theories about monies fuel our economic policies, which usually affect tax structures as well as interest rates. The conventional wisdom of one administration is often replaced by a "better idea" by the next. A relatively recent example of industry restructuring can be found as Norman Augustine challenged conventional wisdom by thinking outside the box. His thesis was that fewer and fewer companies were going to be able to compete for government contracts, especially in relation to the DoD. As CEO of Martin Marietta, he undertook a very aggressive acquisition and merger program that ultimately led to the Lockheed Martin Company, a definite leader in government business. Other companies followed suit (e.g., Northrop Grumman, Raytheon), leading to no less than a restructuring of the industry. Who knows where that will lead as new market forces show themselves.

7.2.5   Government Regulation and Deregulation

We had a rather complex and regulated airline industry in the United States back in 1978. The Civil Aeronautics Board was in charge of "routes and rates" and the conventional wisdom was that this was the natural state of the world (as well as civil aviation). Then, and suddenly to many, we deregulated this industry, trashing the conventional wisdom. The White House and the Department of Transportation led the charge to help bring costs down for the consumer and force new industry efficiencies. Many companies believe that this has worked, but several airlines (some forced into bankruptcy) have opposite views. Is it now time to challenge today's conventional wisdom?

7.2.6   Competition

Successful competition will cause companies to challenge conventional wisdom, at times in the blink of an eye. Other companies have trouble seeing what is happening in the marketplace, and are slow to react. A few that have had their difficulties include IBM, DEC, Wang Laboratories, and Apollo Computer. New competitors continue to be created that are willing to challenge conventional wisdom and try another approach. Just stop for a moment to think about Federal Express, which just pure and simply thought outside the box. Holding on to what's inside the box may well transform your company from a leader into a follower.

7.2.7   Building Complex Systems

Two domains that help us in building and managing complex systems tend to be systems engineering and project management. Both continue to make advances by challenging conventional wisdom. An example is the SoSE construction, discussed in both Chapters 2 and 5. The latter benefits from new ways to measure status and progress, building on earlier methods such as earned-value analysis [7.13].

The bottom line is simply that there are numerous forces at work, all the time, that suggest that a new approach may be called for. Those enterprises that are constantly looking for a better way, in the aggregate, bring about a better approach and solution. Those that blindly follow yesterday's conventional wisdom can be in for a rude awakening.

7.3 MORE CHALLENGEABLE CONVENTIONAL WISDOM

The fine art of management contains a variety of areas in which the challenge of conventional wisdom would appear to be appropriate, although many will question such a challenge. Below we list half a dozen such areas, followed by a brief discussion of each:

1. Stick to what you know how to do.
2. Hire the smartest people for a difficult new project.
3. Smart folks care only about doing good work.
4. Accept what you are told as literally and precisely correct.
5. Accept fully that the correct process will always lead to the correct product.
6. Adopt a leadership style that is a good match to the situation at hand.

7.3.1 Sticking to What You Know How to Do

Peters and Waterman reinforced this piece of conventional wisdom when they said that one of the "attributes of excellence" was to "stick to your knitting" [7.7]. However, out-of-the-box thinking, at least in relation to that approach, is to look for ways to expand what you know and do when new opportunities arise. Further, since new opportunities are almost always present, at least to some degree, challenging this particular piece of conventional wisdom is highly recommended. Of course, part of it is a juggling act so that we keep in balance the investments that we need to make in the old vs. the new. Just about all companies struggle with this issue in a most serious way, just about all the time.

7.3.2 Hiring Smart Folks

We may state some current conventional wisdom: Hire the smartest people you can find to fully staff a new and complex program. Is that the way to go, or is there a built-in trap that needs to be questioned? A new and complex program requires a company's best efforts. Usually, such efforts involve staffing with a group of your best people that have (1) proved their competence on previous *programs in your company*, and have (2) demonstrated that they know how to function effectively and efficiently *as a team*. Teams normally cannot be built overnight. Hiring 10 new people to staff an important and difficult project can be a formula for disaster in that they are unlikely to operate as a true team. Further, the smarter they are, the more trouble you may be in since they will often vie, perhaps indefinitely, for who is the smartest and who has the best idea. This is not conducive to making progress, especially on a difficult project or program.

7.3.3 Smart Folks Care Only About Good Work

Smart folks certainly do care a whole lot about doing good work, but it is by no means the *only* thing they care about. Even the top 1% of performers are fully aware of the economics around them, and to assume otherwise is to make a big mistake. This is not to revoke or support Maslow's hierarchy of needs but simply to recognize that smart folks are paying attention and should be treated with great respect in all aspects of their relationship with your company. That includes salary, bonuses, and perks.

7.3.4 Accept All That You Are Told

The art of communication has many dimensions. In discussions between manager and subordinates, the latter will usually try to place themselves in the best possible light. Various types of "spin" can be put on status reporting, and some important information can be left out. For at least these reasons, the manager generally needs to be proactive in a review situation and ask questions for purposes of clarification and to ferret out the truth. This helps to assure that there is no misunderstanding of what is actually going on, as well as its importance. The declaration that there is "no problem" may not be good enough.

7.3.5 The Correct Process Will Always Lead to the Correct Product

Much attention has been focused recently on making sure that the process is correct. Two examples of this attention are the fields of business process reengineering [7.4] and the capability maturity models [7.13] supported by the DoD and various parts of the government. If the process is wrong, there is essentially no chance that the product will be okay. However, a correct process does not *guarantee* the correctness of the product. The potential missing ingredient is depth of subject matter knowledge in the domain in which the process is defined. The correct software development process applied by a senior software engineer is likely to lead to positive results. The same software process applied by a nuclear engineer who knows very little about software is likely to fail.

7.3.6 Modify Your Leadership Style to the Situation

There is perhaps no argument with the notion that a leader needs to take special note of each particular situation to develop the appropriate response and solution. However, problem definitions and solutions can be largely independent of the style of the leader. If the conventional wisdom is that the leader's style should change as a function of the situation [7.14], questioning that wisdom would be to say that if leaders have a distinctive style (which they usually do) and they have been successful with that style, they probably should stick with it. People tend to do best when they are "being themselves."

7.4 SUMMARY: TWO MEETINGS

It was April of the year 1955. Discussions of how to work together went back to January 1949, instigated by Tom himself. But now it was some six years later and the stakes had become greater. So Jim, Joe, and Sol met again to explore the possibilities.

Jim wanted an exclusive license for his company (X) to produce machines based on the patents held by company Y, of which Joe was the president. Of course, sizable royalties would accrue to company Y, based on the sales produced by company X. Joe and Sol asked for a few minutes so they could discuss the situation privately. Jim said that would be fine with him.

Joe and Sol examined lots of possibilities in a very short time, together with risks, costs, rewards, obligations, expectations, as well as all the sweat equity everyone had already put into the very large project.

They returned to the meeting shortly and informed Jim that they had to refuse to grant an exclusive license but would be prepared to offer a nonexclusive license. Jim said he was not able to accept anything but an exclusive one. So the meeting ended amicably and Jim returned to his company to report on the results to his boss. Who was his boss? Thomas Watson, Jr., and Jim was James Birkenstock of IBM (company X). Joe and Sol were Joe Wilson and Sol Linowitz of Haloid Xerox (company Y).

Some years later, IBM asked A. D. Little to study for them how many large copiers would be needed by U.S. companies, and the answer came back—only a few thousand. IBM shared these results with Xerox, trying to demonstrate that Xerox would be better off partnering with IBM. Joe and Sol conferred again, and confirmed the course they had previously chosen—to bet the company on their ability to bring Xerox copiers to the marketplace on a large and profitable scale.

Conventional wisdom would say this would be too risky and that a teaming arrangement of the type suggested by IBM would be a better solution. This was reinforced by the A. D. Little report, which said the market for large copiers could be very soft. But Joe and Sol questioned all of that conventional wisdom and wound up bringing the Xerox Corporation into full bloom as one of the largest and most profitable companies in the world.

REFERENCES

7.1 U.S. Department of Defense (2003). *Operation of the Defense Acquisition System*, Instruction 5000.2. Washington, DC: DoD, May 12.

7.2 Eisner, H. (1997) The Developer-Off-the-Shelf (DOTSS) Approach, A Progress Report, presented at the Portland International Conference on Management of Engineering and Technology (PICMET), Portland, OR, July 27–31.

7.3 U.S. Department of Defense (1991). *Systems Engineering*, Military Standard 499B. Washington, DC: DoD.

7.4 Hammer, M., and J. Champy (1993). *Reengineering the Corporation*. New York: HarperCollins.

7.5 Mondy, R. W., and S. Premeux (1993). *Management: Concepts, Practices and Skills*. Boston: Allyn&Bacon, p. 154.

7.6 Walton, M. (1986). *The Deming Management Method*. New York: Perigee Books.

7.7   Peters, T., and R. Waterman, Jr. (1982). *In Search of Excellence*. New York: Warner Books.

7.8   Ouchi, W. (1981). *Theory Z*. New York: Avon Books.

7.9   Pascale, R., and A. Athos (1981). *The Art of Japanese Management*. New York: Warner Books.

7.10   Kaplan, R., and D. Norton (1996). *The Balanced Scorecard*. Boston: Harvard Business School Press.

7.11   Goleman, D. (1995). *Emotional Intelligence*. New York: Bantam Books.

7.12   Senge, P. (1990). *The Fifth Discipline*. New York: Doubleday Currency.

7.13   Eisner, H. (2002). *Essentials of Project and Systems Engineering Management*, 2nd ed. New York: Wiley.

7.14   Hersey, P., and K. Blanchard (1977). *Management of Organizational Behavior Utilizing Group Resources*, 3rd ed. Englewood Cliffs, NJ: Prentice-Hall.

Chapter **8**

Perspective 4: Back of the Envelope

Perspective 4—back of the envelope (BOTE)—deals with problem solving by focusing on the most important facts and variables. By doing so, it may be possible to literally sketch out a proposed solution "on the back of an envelope." This approach requires a certain amount of integrative thinking and experience as well as comfort with the use of one's intuition. It also supports the notion that in seeking a solution to a knotty problem, it's okay to start out your thinking with "what if we tried something new and different, like ...?" We presume that such an approach in April 1973 was adopted by some folks who were considering how to start a new package and letter delivery service. There are lots of complicated ways to get packages from one city to another, both overnight and economically. But their BOTE approach was simply to move everything into a central hub by air, and then fan out from there to everywhere else in the country. A bold idea, and easily expressible on the back of an envelope. But would it work? The answer to that lies in just two words—Federal Express. Like another company whose name is synonymous with what they do (Xerox), we don't just send a package from New York to Washington, we *FedEx* it!

Completing this introduction, we also pay homage to and reiterate another idea that supports the BOTE approach. That idea is often expressed as an admonition—K.I.S.S., which means "keep it simple, stupid." Even Eb Rechtin, one of our best systems engineers, reminded us of this necessity in his classic book on the architecting of systems [8.1]. You may have gotten the same suggestion from your boss, from time to time, over the years. Perhaps you've even given it to others around you, speaking from your own experience.

Managing Complex Systems: Thinking Outside the Box,   By Howard Eisner
Copyright © 2005 John Wiley & Sons, Inc.

8.1   WHAT CAN FIT ON THE BACK OF AN ENVELOPE?

It is not difficult to demonstrate that a large number of seminal ideas can fit on the back of an envelope, with room to spare. Here are some examples:

1. Senge's articulation of his five disciplines for a learning organization [8.2]
2. Peters and Waterman's eight attributes of excellent companies [8.3]
3. Covey's seven habits of highly effective people [8.4]
4. Eisner's five key attributes of a leader [8.5]
5. Deming's 14 points for the management of quality [8.6]
6. Byham and Cox's four steps to achieve empowerment [8.7]
7. Kouzes and Posner's six disciplines of leadership [8.8]
8. Drucker's five deadly business sins [8.9]
9. The basic plan–do–check–act (PDCA) cycle

Indeed, the notion of a small number (under 10) of key ideas is supported by a well-known psychologist, G. A. Miller, in a 1956 paper [8.10]. His basic argument is that people can remember about seven things, plus or minus two. That certainly is in consonance with our basic theme and gives us mere mortals some hope for the future.

8.2   SOME EXAMPLES OF BACK-OF-THE-ENVELOPE RESULTS

8.2.1   Strategic Initiatives for My Company

My first example relates directly to G. A. Miller's suggestion of "seven, plus or minus two." In the 1980s, I found myself running an $82 million company that itself was a subsidiary of a larger company. In my first year in this position, I was anxious to tackle the matter of establishing strategic directions for the enterprise. This meant lots of discussion and the construction of a formal strategic plan. I called several meetings of my key vice presidents, attempting to elicit ideas for important strategies that we needed to embrace. As I recall both the process and its results, we wound up with a first list of some 30 strategic initiatives. I sat with my most trusted and ecumenical vice president and we managed to reduce the list to about 15 initiatives. After a few days of further rumination, I pared the list down to 13 initiatives. From that point I felt comfortable asking my vice presidents to put some serious text to these initiatives, which they did. We were finally ready to unveil this opus to my boss, who was the executive vice president of the parent company. As it turned out, he must have somehow internalized Miller's "seven, plus or minus two," and his (almost) immediate reaction was—too many. I asked him how he had come to that conclusion, and his answer was, as I recall, 35 years of experience. "For a company of this size," he explained, "I don't think you can focus on and support, with real resources, that many new initiatives." I accepted his advice and went back to the drawing board with my VPs. When we were finished, the new strategic initiatives numbered less than 10. The actual number perhaps is not the crucial piece of data here; we were certainly very close to Miller's seven, plus or minus two. I think my VPs were relieved to have fewer things to truly worry about and be measured by. It felt more comfortable and achievable for all of us, and we learned a bit more about the true meaning of K.I.S.S. And by the way, I'm sure all the final strategic initiatives could fit on the back of an envelope.

8.2.2 Cost-Estimating Relationships for Software

We know that short lists of linear ideas can fit comfortably on the back of an envelope, as illustrated by the last couple of pages of discussion. But what about a more complex set of ideas, those that are governed by several quantitative relationships? We can also tackle this type of problem, as can be demonstrated by B. Boehm's initial approach to defining a cost estimating relationship (CER) for software. In a landmark book [8.11], an approach called COCOMO (constructive cost model) was introduced and largely accepted by a considerable number of systems and software engineers. The seminal idea was that it was indeed possible to estimate the cost of software by means of a relatively simple formula:

$$PM = A(KDSI)^B$$

where PM is the effort in person-months, KDSI the number of delivered source instructions, and both A and B are constants that can be related to effort multipliers (A) and economies (or diseconomies) of scale (B). In the initial version of COCOMO, Boehm defined modes of operation of the software development effort, and each mode had different values for A and B. For the "organic" mode, where a small team with a great deal of experience is being used, the two key formulas were

$$PM = 2.4(KDSI)^{1.05}$$
$$TDEV = 2.5(PM)^{0.38}$$

where TDEV is the suggested development time. In addition to these equations, COCOMO I introduced two additional variables, as follows:

$$Productivity = \frac{DSI}{PM}$$

$$FTES = \frac{PM}{TDEV}$$

where FTES is the full-time equivalent staff.

Moving forward into COCOMO II, Boehm decided to keep the fundamental form of the effort equation, as shown above [8.12]. However, he expanded and made more explicit the ways in which the values of A and B are modified. In particular, A becomes a function of 7 or 17 variables, and B depends upon 5 variables. Thus, the basic structure of COCOMO II is still manageable in terms of the number of basic variables to be considered.

8.2.3   Certain Types of Warfare

There are few endeavors that are more complicated than warfare. In today's world, the forces are complex and multidimensional, and the weapons that we are able to employ are powerful and can be launched at long stand-off ranges. A "standard" approach to this situation is to employ quite complex simulations as a way of trying to predict the outcome of this type of warfare (see also Chapter 6).

During World War I, however, a British analyst by the name of Lanchester studied combat in warfare and came up with relatively simple results that were used by many, all around the world, for many years [8.13]. Indeed, he produced the *Lanchester equations*, which applied to certain classes of warfare and introduced the notion that combat could be studied fruitfully with an analytic (vs. simulation) approach. That approach dealt with such variables as concentrations of opposing forces (Red vs. Blue), exchange ratios, and superiority in terms of both numbers and weapons. He also came up with a linear law as well as a square law representation, the latter being more complicated, as one might expect. The conception of the Lanchester equations could be described as a set of differential equations that could indeed be written on the back of an envelope. Imagine—dealing with a key aspect of warfare on the back of an envelope! With respect to the solutions to these equations, one had to leave behind the envelopes, whatever their size. But the original concept, remarkably, fits rather well into this category of thinking outside the box.

8.2.4   Views of Architectures

The architecting of new systems, or systems of systems (SoSs), has become a critical topic, especially as our systems grow larger and more complex (see also Chapter 2). I have spent a considerable amount of effort developing and describing a workable and prescriptive architecting process [8.14, 8.15]. As part of that effort, attention has been paid particularly to a *minimum* set of products that will reveal the process as well as important views of the architecture. Implicit in this is an explicit method for architecting the system. This minimum set of products is:

1. A *synthesis* chart that shows the design choices that have been made, for a minimum of three system alternatives, for each system function and sub-function.
2. An *analysis* chart that explicitly evaluates the three system alternatives against a set of formal evaluation criteria, leading to scores for the three alternatives.
3. A *cost-effectiveness* plot that places results of step 2 into a graphical format.

The precise reasons for this approach are described in some detail in the references. The point to be made in regard to the back-of-the-envelope approach of this chapter is that it would take about three envelopes to show the results of the steps suggested above. This takes a hard problem, architecting a system, and condenses it to its short-form fundamentals.

The Department of Defense has tackled the architecting problem in the form of the C4ISR Architectural Framework [8.16]. The first level of results can easily be described on the back of an envelope, and consists of three *views* of an architecture: (1) the *operational view*, (2) the *systems view*, and (3) the *technical view*.

More detailed definitions are:

1. The *operational architecture* view is a description of the tasks and activities, operational elements, and information flows required to accomplish or support a military operation.
2. The *systems architecture* view is a description, including graphics, of systems and interconnections providing for, or supporting, military functions.
3. The *technical architecture* view is the minimal set of rules governing the arrangement, interaction, and interdependence of system parts or elements, whose purpose is to ensure that a conformant system satisfies a specified set of requirements.

For both approaches briefly described above, it is possible to boil the answers down to what would fit literally on the back of three envelopes, attempting to capture the *essence* of the architectures under investigation.

8.2.5 Short-Form Definitions of Systems Engineering

Architecting a system, as described briefly above, is a critical part of systems engineering. However, it is still a subset of the overall process of systems engineering. One question that might arise, in this connection, is: Is systems engineering itself describable on the back of an envelope?

Two one-line definitions of systems engineering were provided in Chapter 2, one from INCOSE [8.17] and the other from me [8.14]. Another short definition is available from Andrew Sage, one of our leading and most prolific authorities with respect to the overall subject of systems engineering, as well as related disciplines: "*systems engineering* is the design, production, and maintenance of trustworthy systems within cost and time constraints [8.18]."

Finally, we should not omit the perspective set forth by the Department of Defense in their Military Standard [8.19], illustrated in Figure 8.1. This very well known chart was developed by the Defense Acquisition University and others and holds a firm place in the history of systems engineering. In this

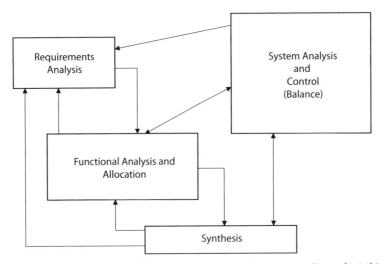

Figure 8.1 Key aspects of the systems engineering process. (From [8.14].)

diagram we note the attention paid to four key elements of the systems engineering process:

1. Requirements analysis
2. Functional analysis/allocation
3. Synthesis
4. System analysis and control (balance)

It is clear that all of the views of systems engineering above are very useful and qualify as sufficiently succinct to fit comfortably on the back of an envelope.

8.2.6 Where We Are Losing Revenues and Profitability

It is expected that many who will be interested in this book will also be looking at business-oriented issues of "making the numbers." That is, you will be looking for some guidance as to how to think about generating higher revenues as well as profits. The first things to focus on if you have an ongoing business whose progress you track from month to month are *revenues* and *profits*. In this respect, you will need a finance and accounting system that produces reliable monthly measurements of these two critical aspects of running a business. The specific aspects of these two measures that you can identify on the back of an envelope are:

1. Revenues achieved last month
2. Profits achieved last month
3. Item 1, by line of business (e.g., by project, for each store)
4. Item 1, for the prior 11 months

5. Item 2, by business area

6. Item 2, for the prior 11 months

That's a total of six critical items. They give you a picture of revenues and profits by *line of business* and also *by time frame*. If you can array revenues and profits by line of business (LOB), you can quickly determine which LOB is making its targets and which are not. You can get a printout of those areas that are okay and those that are not. The picture is made more complete by looking at results plotted on a time line for this month as well as the preceding 11 months. Graphs as a function of time give us an "eyeball" way to see first derivatives that establish trends, both good and bad. All of this tells you basically where you're having trouble and where you're not, more or less at a glance. It also sets the stage for asking a lot more questions about "why": that is, why project 7 fell short, and why store 58 is not operating up to snuff. In other words, a very quick assessment of status can literally be made on the back of an envelope. Finding out why and updating projections for future months usually require a deeper look that transcends this single-envelope approach.

8.2.7 Great Ideas in a Sentence or Two

Many ideas are deceptively succinct in their expression but wind up being very powerful in their implications. Indeed, the final idea can be expressed on the back of an envelope even though lots of thought may have gone into them. Here are some of these ideas that are recognizable by most people. As you will see, they can fit on the back of an envelope and they have also given rise to much progress and contributions by others.

The earth is not the center of our solar system. One of our early conjectures, but a "theory" that many thought was correct and that it was sacrilegious to think otherwise. Thank goodness we got this straight before we started sending rockets everywhere.

 The earth is round rather than flat. This idea is only seven words but may have had something to do with the discovery of both North and South America. We're told this was obvious if you looked carefully at ships as they approached from the ocean. Hindsight is still wonderful.

It's not possible for a physical entity to travel faster than the speed of light. Of course, one of Einstein's major contributions. It was indeed difficult to conceive of this "discontinuity" or barrier in the middle of our physical space and systems. It's now about 100 years later and we've gotten quite used to the idea. But it's still only about 15 words long.

$E = mc^2$. In this very simple equation, Einstein told us what the transformation was between mass and energy. The c-squared multiplier is quite a large number and had something to do with understanding how much energy could be released from an atomic bomb.

Maxwell's equations. In just four equations, clearly no larger in width or girth than the back of an envelope, James Clerk Maxwell set the stage for understanding much of the foundation of electric and magnetic field propagation. The electro-magnetic spectrum, and its applications, is everywhere in our lives.

We can send information by changing the frequency of a radio wave. Maxwell told us what radio waves were and how they functioned, but others chimed in on the world of applications. This particular one has to do with how frequency modulation (FM) was conceived.

Newton's gravitational law. Of course, Newton was one of our great geniuses, and the gravitational law was only one of his many contributions. This inverse square law fits on a small envelope but tells quite a story.

Information-theoretic entropy. Claude Shannon, an engineer who clearly thought outside the box, conceived of a formula for entropy that established the basis for the field of information theory. This, in turn, provided insights that led to code breaking and the formulation of a variety of coding techniques. Other folks connected this notion of information-theoretic entropy to thermodynamic entropy.

Man can fly, but not like a bumblebee. The reason for including this "negative" idea is to recognize that certain contributions are made in the negative. By so doing, they exclude an entire array of approaches that probably would not bear fruit, thus saving the time and effort of lots of potential investigators. Of course, we have indeed learned how to fly a different way from the way that birds and bees do, which could also be thought of as clearly thinking outside the box (conventional wisdom being to try doing it the way birds and bees do it).

The point of discussing these factors is simply to say that certain important and far-ranging insights have literally been expressible on the back of an envelope. This is not to minimize the effort that went into developing those insights. But many powerful answers to important questions can be stated in very succinct terms, and we should try to look for such answers as part of this chapter's approach to thinking outside the box.

8.3 CONSTRUCTING WHAT'S ON THE BACK OF THE ENVELOPE

Articulating the seminal idea on the back of an envelope is probably not a problem if you're a genius, but what about the rest of us who are shooting for the top 5% instead of the top 0.000005%? Let's take a look at some of the steps that might be helpful in approaching this matter, such as the following:

1. What is the problem?
2. What are the key factors, variables, and observables?
3. What inferences made be drawn from step 2?
4. What are some alternative solutions to the problem?
5. What appears to be the best solution?

A few examples of how one might use these questions follow after the brief discussion below:

- **What is the problem?** The necessity to state the problem has been well recognized and established. Try to do it in just one sentence, if possible. Here again, in this BOTE perspective, we are trying to be as terse as possible. We want to address one problem, not several. Use the imperative K.I.S.S. to the maximum extent possible.
- **Key factors?** Just write down what comes to mind, without any explanation. A linear list will do. If you can connect any of these factors using a diagram, that's fine.
- **Inferences?** In looking at each of the factors, variables, and observables, do you see any inferences that can be drawn? If so, write them down. If not, don't press the point.
- **Alternative solutions?** Here you're thinking about not one, but several solutions. Try three solutions, if you can, one sentence for each. More than one solution helps to broaden your field of view, and hopefully increases the chances that a truly preferred answer will emerge.
- **Best solution?** Pick, from among the alternatives above, the one that seems the best. Write it down. Also write down why you think you have the best answer. Think it over a second or third time to see if your selection changes.

Now let's work our way through a few examples of the use of these five steps suggested.

8.3.1 Landing on the Moon

Some while ago we were facing the problem of how to send a person to the moon and back. Let's try our five-step approach with respect to that matter.

- **Problem?** To transport a person successfully from Earth to the moon and back.
- **Key factors?** Thrust required, total fuel needed, safety, budget, schedule, probability of success.
- **Inferences?** Fuel, safety, and probability of success are critical.
- **Alternatives?** Seem to be three: (1) a "direct" approach, which involves the following steps—Earth to Earth orbit to moon to Earth; (2) an "Earth orbit meet" approach, which involves Earth to Earth orbit to moon to Earth orbit to Earth; (3) a "moon orbit meet" approach, which involves Earth to Earth orbit to moon orbit to moon to moon orbit to Earth.
- **Best solution?** Select alternative 3.

A footnote to this problem is that NASA, after studying the three alternatives above, eventually selected the third alternative, which apparently was deemed to be almost $1.5 billion less expensive and achievable on a six- to eight-month shorter schedule [8.20]. Safety was considered to be a "wash" among the three possibilities.

8.3.2   Project Management Problem

Now let's take a brief look at a possible situation on a project and what the project manager might do about it [8.14]:

- **Problem?** The project is overspent by 13%; the lead systems and software engineers fight with each other all the time; there's a project review session scheduled with the customer the day after tomorrow.
- **Key factors?** Cost vs. budget; schedule; engineers fighting; customer information and acceptance.
- **Inferences?** The project is in trouble and some type of action is required.
- **Alternatives?** At least yes vs. no in following areas: (1) hold staff meeting; (2) determine cost status; (3) determine schedule status; (4) determine technical progress status; (5) talk privately to both engineers who are having conflict; (6) cancel customer meeting.
- **Best solution?** Selected (1) yes, immediately; (2) yes, (3) yes, (4) yes, (5) no, (6) yes; but reschedule two days later.

The project manager knows there is a serious problem and wishes to see how deep the problem is and explore solutions. The conflict between the two engineers may turn out to be serious, but dealing with it can be delayed. There's not enough time to investigate the problem and find a solution before the scheduled meeting with the customer, so that meeting is put off for a couple of days.

8.3.3   How Can I Plan My Career?

George has been working on radar systems since he got his bachelor's degree in electrical engineering six years ago. He's been offered a new opportunity in the same company to work on air defense systems. He views this as a major step, one way or another, in his career. He wants to try the BOTE approach to thinking about the problem.

- **Problem?** Should I broaden my view of engineering systems, or deepen it by becoming even more expert in radar systems?
- **Key factors?** Personal interest in broader or deeper picture; job opportunities; pay now and into the future; personal interest in becoming a manager down the road; timetables.
- **Inferences?** The choice that is made now may be a critical step in establishing a career direction.
- **Alternatives?** (1) accept position to work on air defense systems; (2) stay with current position on radar systems; (3) try to negotiate for a one-year assignment on air defense systems, with an option to return to radar work after a year.

- **Best solution?** George selects solution (1). He wants to try broadening his perspective about systems to see if that approach is satisfying. He does not select (3), reasoning that such an option is probably open to him anyway. He also knows there is some risk to not selecting (3) if he is truly uncertain about which way to go at this time.

8.4   WHAT DOES IT ALL MEAN?

This perspective of problem solving on the back of an envelope is basically a metaphor for how to approach a new way of thinking. When confronted with what appears to be a difficult or complex problem, many people begin by shredding the problem into many pieces and then shredding the pieces into even smaller pieces. We are encouraged, to some extent, to do exactly that by the functional decomposition element of systems engineering. However, the danger is that in trying to get some sense of the forest as a whole, we wind up looking at each and every tree. Then we are unable to put the pieces back together. The former is an analysis process and the latter a synthesis process.

The primary task represented by the BOTE approach is to *simplify rather than make more complicated.* We are trying to explore the *essence* of the problem and its solution. The process is largely a synthesis, where we attempt to see through the complexity to the basics and fundamentals. This is difficult to do, and there is some error that goes along with such an approach. But there is usually an opportunity to go back and reexamine the issues and change your mind when you are given more time to examine the problem. The process is especially useful for the fine art of proposal writing because when a team is writing a proposal, there is usually not much time, and answers have to be developed quickly. Hopefully, if this approach can be mastered, the team will win the contract and get another chance to put more effort into working all aspects of the problem.

8.5   SUMMARY: A MEETING

John, a senior-level manager in the ABC Company, has been asked by his boss to reengineer his operations so as to achieve a 10% increase in productivity in the next four months. John and his boss agree that cost reductions are not necessarily in order if, for example, the 10% productivity increase is achieved at the same cost as the cost at present.

John immediately calls a meeting with the eight key managers who report to him, each of whom is handling a discrete part of the overall operation. He also invites his administrative assistant to take notes as ideas are presented. John facilitates a lively discussion by soliciting ideas as to how to achieve the productivity goal. Considerable attention is paid by the group to business process reengineering (BPR). As the two-hour meeting is brought to a close, John asks everyone to think about the problem and be prepared to explore solutions over an internal business lunch the next day.

At noon the next day the team reassembles and John goes around the room asking for approaches toward a solution. A dominant theme emerges: that of hiring a consulting firm to come in and carry out a BPR assignment. The pros and cons of this approach are discussed and there is clear support for moving in this direction. Only a couple of managers have read Hammer and Champy's landmark book on BPR [8.21], and both of them are strong supporters of this approach.

About an hour and a half into the discussion that appears to be converging toward the approach noted above, one of the managers (George) begins to articulate his views of the matter. Essentially, he starts by focusing on two issues:

1. The apparent fact that only about one-third of the BPR efforts carried out in the recommended fashion appear to have been successful (two-thirds have been viewed as failures, in varying degrees).
2. The apparent fact that various consultants/experts appear to want a minimum of $500,000 to come up with the reengineered solutions.

George sees these facts as serious obstacles to success and begins to place another alternative on the table, having thought at least to some extent about cost, risk, and empirical data that he accessed over the Internet in the past 24 hours, the precepts of Hammer and Champy, and the overall productivity goal. He describes his recommended approach as "skimming the cream" (STC) and his belief that getting a 10% improvement in productivity was achievable by such an approach. Further, his belief extended to the idea that each manager could indeed find a 10% improvement (approximately) by looking at the problem with his or her own people and without the use of a consultant. Slowly, George's thinking-outside-the-box approach gained supporters until it became the approach selected by John. The STC solution was wildly successful, as the group depended on its own skills and goodwill rather than that of a consultant. The overall task was ultimately completed ahead of schedule (in three months instead of four) for about $50,000, one-tenth of the more conventional BPR approach. George's questioning of conventional wisdom allowed him to explore a solution outside the box, and the others saw and accepted the wisdom of his approach. Over a year's worth of measurement after implementation, productivity was increased by approximately 11%.

REFERENCES

8.1   Rechtin, E. (1991). *Systems Architecting*. Englewood Cliffs, NJ: Prentice Hall.

8.2   Senge, P. (1990). *The Fifth Discipline*. New York: Doubleday Currency.

8.3   Peters, T., and R. Waterman, Jr. (1982). *In Search of Excellence*. New York: Harper & Row.

8.4   Covey, S. (1989). *The Seven Habits of Highly Effective People*. New York: Fireside/ Simon & Schuster.

8.5 Eisner, H. (2000). *Reengineering Yourself and Your Company*. Norwood, MA: Artech House.

8.6 Walton M. (1986). *The Deming Management Method*. New York: Perigee Books.

8.7 Byham, W., and J. Cox (1988). *Zapp! The Lightning of Empowerment*. New York: Harmony Books.

8.8 Kouzes, J., and B. Posner (1993). *Credibility*. San Francisco: Jossey-Bass.

8.9 Drucker, P. (1995). *Managing in a Time of Great Change*. New York: Truman Talley Books/Plume.

8.10 Miller, G. A. (1996). The Magical Number Seven, Plus or Minus Two: Some Limits on Our Capacity for Processing Information, *Psychological Review*, Vol. 63, No. 2, March pp. 81–96.

8.11 Boehm, B. (1981). *Software Engineering Economics*. Englewood Cliffs, NJ: Prentice-Hall.

8.12 Boehm, B., et al. (2000). *Software Cost Estimation with COCOMO II*. Upper Saddle River, NJ: Prentice-Hall.

8.13 Morse, P., and G. Kimball (1952). *Methods of Operations Research*. Cambridge, MA: Technology Press of Massachusetts Institute of Technology; New York: Wiley.

8.14 Eisner, H. (2002). *Essentials of Project and Systems Engineering Management*, 2nd ed. New York: Wiley.

8.15 Eisner, H. (2003). *Eisner's Architecting Method (EAM): Prescriptive Process and Products*, presented at the 13th Annual International Symposium, INCOSE, June 29–July 3, Washington, DC.

8.16 U.S. Department of Defense (1997). *C4ISR Architectural Framework*. v. 2.0. Washington, DC: Architectures Working Group, DoD, December 18; this framework has been renamed the *DoD Architectural Framework*, or DoDAF.

8.17 See www.incose.org.

8.18 Sage, A. (1992). *Systems Engineering*. New York: Wiley.

8.19 U.S. Department of Defense (1991). *Systems Engineering*, Military Standard 499B. Washington, DC: DoD.

8.20 Buede, D. (2000). *The Engineering Design of Systems*. New York: Wiley.

8.21 Hammer, M., and J. Champy (1993). *Reengineering the Corporation*. New York: HarperCollins.

Chapter **9**

Perspective 5: Expanding the Dimensions

When we think about dimensions, we often go to the three spatial dimensions and the one time dimension. We are used to and live in these four dimensions, and happily so. There are times, however, when our best thinkers are telling us that these dimensions can and should be expanded in order to answer some pressing questions of the day. For example, in tackling the matter of the grand unified theory (GUT) of physics, we are being told that answers might lie in 10-dimensional thinking. Even though we cannot visualize 10-dimensional space, our mathematical solutions to the GUT problem have moved us to that precarious perch. But more about that later in this chapter. Suffice it to say now that if we could *definitively* formulate the GUT, otherwise also known as the theory of everything, in 10 dimensions, the journey would be well worth it, on its own as well as where it might point us for the future.

To come down to a more mundane but very practical level of dimensions, when we look at a spreadsheet printout we are often looking at only two dimensions of a problem. Although we have the capability to develop a three-dimensional spreadsheet and to see how it might change over time, we often do not do so. To the extent that this is true, we are choosing to stay inside the box. When we add the third and fourth dimensions (one in space, the other in time), we are allowing ourselves to venture outside the box.

Yet another example of this perspective regarding thinking outside the box is the following match problem. We see in Figure 9.1 a total of six matches placed on a desk, in a particular formation. The problem is to move three of these matches so as to make four equilateral triangles. This is a difficult problem to ponder, and you will have the rest of this chapter to do so. A solution is provided later in the chapter.

Managing Complex Systems: Thinking Outside the Box,   By Howard Eisner
Copyright © 2005 John Wiley & Sons, Inc.

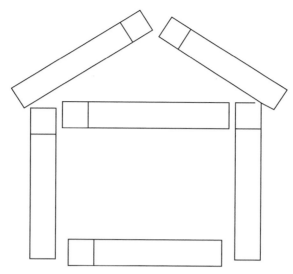

Figure 9.1 Match replacement problem: move three matches to form an overall figure with four equilateral triangles.

As we explore various ways of expanding dimensions in this chapter, we may also notice that this way of thinking might be considered a "first cousin" to that suggested in Chapter 5 dealing with broadening and generalizing. To this author, however, there were and are enough interesting avenues to explore to treat these two ways of thinking as separate for the purposes and goals of this book.

9.1 ANOTHER LOOK AT ARCHITECTING

In earlier chapters, notably Chapters 5 and 8, the notion of architecting systems was discussed. Architecting is a key element of systems engineering and is considered to be a top-level design activity that sets the stage for success or failure of the system. The question to be addressed in the context of this thinking perspective— expanding the dimensions—is: How can the architecting process be supported (if at all) by an expansion of the dimensions of the problem?

The specific architecting procedure that I set forth is based explicitly on the cost-effectiveness consideration of alternatives. It is the construction of these alternatives that specifically expands the dimensions. It was shown in Figure 5.1 that the space of system solutions contains at least the following three points:

1. A low-cost alternative that nominally satisfies all system requirements
2. A knee-of-the-curve alternative that yields especially high increments in effectiveness per unit of cost
3. A high-effectiveness alternative whose cost is also likely to be high

This arhitecting process thus has explicitly expanded the dimensions so that (at least) three alternatives are considered rather than just one. This is a good example of how dimensions can be expanded for the purpose of obtaining a better architecting solution. More details regarding this method of architecting are explored in Chapter 13, which deals directly with the systems approach.

9.2   THE MORPHOLOGICAL BOX

Within the architecting process is a synthesis step whereby alternative design approaches are considered for each functional area of the system. The broad question in this connection is: How can we assure that all reasonable design alternatives have been defined and considered in the architecting of a system?

The general answer to this question is that 100% assurance is not possible. Basic limits in our ability to conceive of all possible solutions will probably always prevail. However, we can try to use methods that will, in effect, force us to consider a broad range of possibilities. One such method, created by F. Zwicky, involves use of the *morphological box* [9.1]. This procedure has proven itself and has been accepted to the point that morphological structuring and analysis is standard fare in some arenas. By its very nature, creating the morphological box increases the dimensions of the solutions to a particular design problem.

Zwicky defines five steps that should be used to carry out his suggested morphological analysis:

1. Formulate the problem in concise terms.
2. Define all important parameters that bear on the problem.
3. Construct a multidimensional matrix that represents the morphological box.
4. Evaluate all the solutions that are part of the morphological box.
5. Select the best and most suitable solution(s).

We note that for step 3, all relevant dimensions are defined and mapped in the form of a matrix. As an example, if we were trying to develop a new automotive power plant, we might start by defining all the various types of fuels you could think of, and mapping them against all the various engine types you could think of. Those that didn't make sense would be rejected, and the others would carry forward into the next step. In some cases the fuel itself will not have an equivalent engine using that fuel. This creates, in principle, the possibility of designing a new engine that will accept that fuel. Various combinations would also be considered so as not to miss hybrid vehicles, which combine two or more "solutions." Clearly, this method can lead to new combinations and ultimately to the invention of systems that are not available as commercial products. Thus, some of the power of Zwicky's approach lies in its ability to conceive of new types of systems. It is also a prime example of how to think outside the box by expanding the dimensions in a very well-defined and explicit way.

9.3 HAVE YOU VISITED FLATLAND?

A classic book by Edwin Abbott, called *Flatland* [9.2], is about inhabitants of a world with a flat surface (i.e., it had only two dimensions). The author takes us through various journeys in two dimensions, giving us an excellent idea of what living was like in such a limited world. In the foreword by Isaac Asimov [9.2], we have the following quote: "Flatland is not just an amusing and witty exercise in geometry, but it is a dissertation that could lead to profound thought about our Universe and ourselves." It is for this reason, and particularly how it relates to expanding the dimensions, that it is included in this book.

Moving ahead with the story: A Sphere appears, a stranger from Spaceland, and begins to convince our main character (MC) that there is such a thing as a third dimension. This is finally demonstrated, and our MC becomes convinced. He then goes back to his friends in Flatland and tries to explain the Gospel of three dimensions. You can probably guess what happened next. Yes, he was thrown into jail, and as he narrates the story, he has spent the last seven years in that precarious place. One of his final comments is his hope that he may be able to find his way to the "minds of humanity in Some Dimension, and may stir up a race of rebels who shall refuse to be confined to limited Dimensionality."

It is easy for us to accept the three-dimensional physical world since we live in it. It was, however, extremely difficult for the MC in Flatland to accept that world. One may assume that such is the case when we attempt to accept a four-, five-, or 10-dimensional construction of the world we live in. A message of *Flatland*, one of many, is that we should be somewhat more open to these ideas, especially when we see our scientists working in these multidimensional constructs.

In 2001, another book appeared that, to some extent, picked up where *Flatland* left off [9.3]. The author called it *Flatterland*, which he said was "like flatland, only moreso." If you've gotten through *Flatland*, you now have a suggested pathway to the next conception incorporating these types of ideas.

9.4 FUNCTIONS OF MANY VARIABLES

In a brief attempt to move the *Flatland* adventure to a more credible place, one need go no further than simply to recognize that we have, for quite a long time, been dealing with functions of many variables. Even though there are limitations as to how many physical dimensions we can easily represent, we have no difficulty "seeing" functions of a large number of variables. There are numerous examples, but let us go back to the basic formulation of COCOMO, discussed in Chapter 8 [9.4]. You will recall that the person-month equation had a simple form:

$$PM = A(KDSI)^B$$

We also discovered in COCOMO II that A was a function of 7 or 17 variables and B was a function of 5 variables. So PM is itself a function of some 12 or 22 variables.

These are not physical dimensions as we know them, but we are clearly dealing with some type of multidimensional "space." This is well within our capacity to readily accept; lots of variables often need to be brought into play to make a top-level calculation. If this is still not convincing, either take a long look at what the Weather Service does for a living, or read a recent book about superstring theory (see Section 9.10).

9.5   THE MOVIE CAMERA

The movie camera can be thought of as a device that came about by an expansion of dimensions. Imagine looking at a snapshot and thinking about how to convert it into something that replicated the natural motion of everyday life. With the benefit of hindsight, a representation over many "time slices" comes to mind. The time-sliced snapshots must each be different, conveying the change of scene as time proceeds. Thus, in going from a still photo to the moving picture, both the photo and the time dimension are being expanded.

All of this became clear in the special insight provided by one of our most creative inventors, Thomas Alva Edison. He lived from 1847 to 1931 and was known as the Wizard of Menlo Park, New Jersey. He is perhaps best known for the incandescent lamp, but in fact he was one of our most prolific inventors, with some 1093 patents to his name. Using results of stop motion photography from an Englishman, he brought the idea farther down the road and introduced the first movie camera in 1890 [9.5, 9.6]. Other information about Edison, one of the best out-of-the-box thinkers, was provided in Chapter 4.

9.6   A MULTIFUNCTIONAL DEVICE

From earlier discussions we can see that expanding the dimensions in looking at an issue or a problem can be carried out in many ways. Physical space and time are our "four dimensions" and are recognizable by everyone. However, other dimensions exist in problem solving, especially in dealing with the expansions of *functionality*. Many issues are addressed in this domain, and they are meant to be included in this discussion of expanding the dimensions.

An example of the above is the case in which we take a unifunctional device and expand the dimensionality of its use by adding one or more new functions. An especially interesting set of developments can be seen by looking at the following four devices:

1. The telephone
2. The camera
3. The personal digital assistant (PDA)
4. The digital computer

As a starting point, each of these can be taken in its elemental form, before it was combined with the others to add new functionality. Now it can be argued that these four devices are being "crunched together" so that consumers can expect to have at least some features of all in one device. This confluence of technology is a testament to both our engineering and our marketing prowess. To have the handheld telephone act as the "platform" to integrate all these functions is to be expected, and we can predict further definition and integration of functions as we go down the road. When we find a good thing in the marketplace, there is apparently no end to the energy we are willing to put forth to make that good thing better.

The bottom line is that adding functionality is another way to think about expanding dimensions, all of which is an example of this way of thinking outside the box. But read on to take another look at how increased functionality may be viewed.

9.7 A MULTIFUNCTIONAL HOUSE

If you read the floor plan for a house you will notice that we usually identify the various rooms with an appropriate adjective. Thus, we have a "living" room, a "dining" room, and "bed" rooms. However, we have all probably had the experience of seeing how friends have converted from one type of room to another: Dining rooms have been converted into offices; bedrooms have become guest rooms, dens, and offices; and living rooms have been changed into family rooms as well as places in which no human being can enter unless they are heavy-duty "company." So, despite the fact that we still use conventional adjectives in describing these rooms, many of us are converting their use to suit our needs. And this conversion continues as our needs change, such as converting a bedroom into a guest room when the oldest child is off to college or moves to another area to begin a working career.

In this case, consumers are being creative in the way they use the rooms of their houses. Perhaps this is a signal to builders to find new ways to design houses to create new opportunities for conversions of the rooms. Can you think of any new ways you would use this principle in respect to the features of the rooms in a house? Does this apply to office space?

9.8 WHERE DO ELEVATORS BELONG?

There are a large number of multistory buildings that do not have an elevator, despite the fact that walking up four stories, for example, can be a distinct hardship for the inhabitants. The knee-jerk reactions to this problem appear to be (1) to look for some interior space that can be used to construct an elevator shaft, or (2) to do nothing. The former is often not workable or is extremely expensive and inconvenient. This, in turn, can lead to the latter choice. These selections, one might conclude, are distinctly inside the box, even when an interior solution is adopted.

On the other hand, some people seem to be thinking outside the box, or in this case, the building. Why not, given the right exterior design, build a set of elevators (or escalators) connected to the *outside* of the building?

Armed with this question, an Internet search revealed that there is at least one company that is prepared to build an exterior elevator. That sounds like a move in the right direction, one that recognizes a special need of people with certain types of disabilities. One might also say that this is a clear expansion of dimensions, specifically those associated with the size structure of a building. Expanding the dimensions of a problem–solution combination certainly can come in many shapes and sizes.

9.9   WHERE ARE AIRPLANES SUPPOSED TO FLY?

For many years, aircraft have been flying on well-defined pathways between terminal areas. These are a type of "highway in the sky," and we have been relatively comfortable with being able to positively control the flow and safety of these airplanes. However, as the traffic increases, these highways have become more congested, resulting in greater delays across the system. To improve this situation, the Federal Aviation Administration has embarked on a "free flight" program that allows aircraft to deviate from these fixed pathways under the right sets of circumstances. This is a clear expansion of the dimensions of the problem, as airspace previously unused will now be usable. The new airspace is massively increasing effective airspace capacity dramatically.

The free-flight notion can only be achieved with the systematic introduction of new equipment and procedures. The program designated as Free Flight Phase 1 began in 1998 and involved the deployment of various new tools, such as [9.7]:

1. A traffic management advisor
2. A user request evaluation tool
3. A surface movement advisor
4. Collaborative decision making

The Phase 1 program was focused on well-defined goals, including:

1. The achievement of early benefits to all stakeholders, using well-known and proven technologies
2. Sustaining or improving safety levels
3. Having core capabilities available by the end of 2002
4. Maintaining consensus throughout the U.S. aviation community
5. Using an evolutionary design and development approach

Free Flight Phase 2 is to expand on the best results of Phase 1 and carry out research that is designed to reduce congestion and provide improved access to the

National Aviation System. So we have a relatively large real-world program moving along that is based on a central idea about expanding the dimensions of air traffic control. This allows flight paths off the prior route designations, thus providing a potential to enhance capacity and at the same time, to maintain or improve safety for all users of the system.

9.10   THE GRAND UNIFIED THEORY

As alluded to at the beginning of this chapter, our best scientists and mathematicians have been in search of the grand unified theory, known to many as the theory of everything. The motivation for this search is very clear: Such a theory will unify physics by unifying the four basic forces:

1. Gravity, a "universal" force
2. The force of electromagnetism
3. The strong nuclear force (binding nuclear particles together)
4. The weak nuclear force

Many have contributed to a better understanding of these forces, from Newton and Einstein with respect to gravity, to James Clerk Maxwell, who formulated the essential equations for electromagnetism in the 1850s. Curiously, writing down Einstein's gravitational field equations in five dimensions (rather than four) led to the usual four-dimensional gravitational equations, along with an additional set of equations that were the same as Maxwell's equations [9.8]. This was found in 1921, an early adventure into the world of more than four dimensions. It also shows how expanding the usual dimensions can lead to something extremely significant and useful.

In the 1970s, two physicists (John Schwarz and Michael Green) led the charge toward developing string theory, and its modern form, known as *superstring theory*. This is a unifying theory in 10 dimensions. As John Schwarz remarked [9.8]: "So the theory will have extra dimensions. . . . The solution of the equations entails curling up these extra six dimensions into a little ball sufficiently small that it wouldn't be observed." Can you relate to this observation, or is it just a bit too far outside the box?

If this 10-dimensional foray is insufficient for your taste, you may appreciate a spin-off known as *membrane theory* [9.9]. This, in turn, has led us to an *eleventh* dimension.

There seems to be little doubt that these theories are moving forward despite the fact that we are still in a process without a definitive and widely accepted result that can be called *the* grand unified theory or *the* theory of everything. All of this, however, does seem to support the notion of trying to expand the dimensions with respect to a problem you're facing. Our primary domain in this book is building and managing complex systems rather than the most fundamental law of physics. But we can also see lots of examples in which we are able to find better solutions if we look for other, perhaps new, dimensions of the problem.

9.11   THE SPREADSHEET REVISITED

Although two-way tables had been with us for many years, it was not until about 1979 that such tables were converted into software for early personal computers (PCs). The genius behind that conversion was Dan Bricklin, a computer application designer and a student at the Harvard Business School. Bricklin was the designer of VisiCalc, and a colleague by the name of Bob Frankston was the programmer who wrote the code [9.10]. The early target machines included the Apple II, TRS-80, and Commodore PET. One might say that this was the first piece of software truly deserving of the title Killer Ap (application). The Killer Ap was the object of the search for the computer application "Holy Grail."

I can remember quite clearly using VisiCalc on my Apple II Plus machine. It was indeed a joy to learn and a lot of fun to use. The concept was nothing less than brilliant, and the implementation wonderful, as the programmed cells rearranged their data with each new input. We've come a long way from VisiCalc to Lotus 1-2-3 to Excel, and today each spreadsheet version incorporates new and more powerful features. One of the clearest examples of expanding the dimension occurred when the spreadsheet went from a two- to a three-dimensional representation. Here's how this expansion might look as one constructs a life-cycle cost model (LCCM) for a system.

First, we develop two dimensions of the LCCM:

1. The rows are the individual cost elements.
2. The columns are the years in which each cost expenditure is experienced.

Three typical top-level cost element categories are:

1. Research, development, test, and evaluation costs
2. Procurement or acquisition costs
3. Operations and maintenance, sometimes also called operations and support costs

These, in turn, may be broken down further into individual cost elements so that a total of some three dozen rows may constitute all of the cost elements [9.11]. Let us assume a useful life for the system of 20 years, so we have a two-dimensional spreadsheet with some 36 rows and 20 columns, yielding 720 total cells.

This two-dimensional version of a LCCM spreadsheet may then be expanded to include a third dimension. The latter is simply the set of subsystems that make up the overall system. Thus, we have expanded the two-dimensional spreadsheet to three dimensions, as depicted in Figure 9.2. In today's world of high-powered

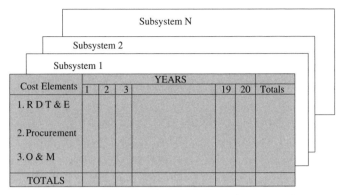

Figure 9.2 Three-dimensional life-cycle cost model.

spreadsheet designs, such conceptions are commonplace and are easy to invoke for real systems. The bottom line is that a system LCCM can be constructed with a three-dimensional spreadsheet with a data-intensive but simple three-dimensional structure. Another example follows in the hypothetical meeting described below.

9.12 SUMMARY: A MEETING

It was September 16 and John, the new president of the JML Corporation, was holding his third monthly measurement meeting for the company. John was in the main conference room with George, the COO, Patricia, VP finance, and Frank, the EVP. Sales and profits had been strong for the months of June and July and John expected the same for August. George started with viewgraphs showing revenues, costs, and profits, by business area, for August. The data were presented in several spreadsheet tabulations, with cost elements as rows and business areas as columns. The profit percentage seemed okay to John, but he thought he remembered higher revenues for prior months. He asked for the results for June and July, and Patricia produced them from her folder. When George projected these numbers on the screen, John read them with both care and alarm. "My off-the-top calculations show that we're almost 20% down from the June–July numbers. What's going on here?", he questioned.

"August is always a bad month," said Frank, trying to soften the question.

"I hear the words," said John, "but I need to see it in black and white."

George, Patricia, and Frank all looked at each other. After a moment, Patricia got up and asked for a five-minute break. "I'll be back soon" she said as she left the room, headed for her office.

She returned, true to her word, in four minutes. "Never throw anything away," she said, as she projected a slide on the screen. It showed revenue dollars and profit percentage for every month of the preceding year. It also confirmed what Frank had said. Revenues were down about 25% in August last year from the numbers in June and July. This August, the profit percentage was better than that for last August.

Matches Flat on a
Surface

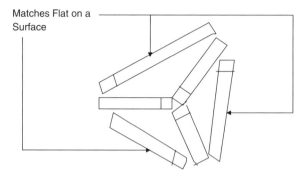

Figure 9.3 Solution to the matchstick problem: perspective drawing of six matches forming four equilateral triangles, in three-dimensional space.

August was always a bad month, they all now remembered, and the "crisis" with John was passing.

"I got it," said John. "Now let's remember to see what we can do to improve the numbers next August. And many thanks for the quick retrieval of the data from last year, which I was obviously not aware of."

This monthly measurements scene involved many pages of spreadsheet tabulations, but one of the dimensions was missing: the data that would provide a look at last year's results month by month. In retrospect, this new look at the time dimension should have been included, especially knowing that the August results would cause John heartburn. It's not likely that these data will be missing from future monthly measurements meetings, and it is likely that new expansions of the data dimensions will be discovered as more meetings occur.

9.13 SOLUTION TO THE MATCHSTICK PROBLEM

A matchstick problem was introduced in Figure 9.1. The problem is to move three of the matches so as to make four equilateral triangles. A solution can be found when you allow yourself to move beyond the two-dimensional space in which the problem was presented. An answer is evident when you move the three matches into the third (vertical) dimension and prop them up as shown in Figure 9.3. All of a sudden, the answer is obvious. If the figure is a bit obscure, try it with six real matches. Using the third dimension is liberating but not an obvious place to go. Many real-world problems are similar—we're stuck until we find a way to look at the larger dimensionality of the matter.

REFERENCES

9.1 Zwicky, F. (1969). *Discovery, Invention, Research*. Toronto, Ontario, Canada: Macmillan.

9.2 Abbott, E. (1884). *Flatland, A Romance of Many Dimensions*. New York: Harper & Row.

9.3   Stewart, I. (2001). *Flatterland.* Cambridge, MA: Perseus Books.

9.4   Boehm, B., et al. (2000). *Software Cost Estimation with COCOMO II.* Upper Saddle River, NJ: Prentice Hall.

9.5   Mannion, J. (2003). *The Everything Great Thinkers Book.* Avon, MA: Adams Media Corporation.

9.6   Balchin, J. (2003). *Science—100 Scientists Who Changed the World.* New York: Enchanted Lion Books.

9.7   http://ffp1.faa.gov.

9.8   Davies, P. C. W., and J. Brown (eds.) (1988). *Superstrings: A Theory of Everything?* Cambridge: Cambridge University Press.

9.9   Morris, R. (1999). *The Universe, The Eleventh Dimension, and Everything.* New York: Four Walls Eight Windows.

9.10   Cringely, R. (1992). *Accidental Empires.* New York: HarperCollins.

9.11   Eisner, H. (2002). *Essentials of Project and Systems Engineering Management.* New York: Wiley.

Chapter **10**

Perspective 6: Obversity

Obversity, perspective 6 for thinking outside the box in regard to building and managing complex systems, will be taken here to be the negative counterpart of an affirmative proposition. In straightforward terms, the obverse statements presented here will be declarations that, if followed, will lead one in the direction of failure. Why would you be interested in reading such statements? The answer is that I have found that people pay more attention to obverse statements than to positive admonitions. Perhaps we are all tired of hearing these admonitions again and again. Whatever the reason, it is hoped that you will suspend judgment until you have had a chance to see exactly what obverse suggestions look like.

Another view of obversity is simply that if you acknowledge a particular obverse statement, you may be inclined to take action in the opposite direction. The reason is quite direct. If you can see that a particular obverse action is likely to lead you down the wrong road, several obverse actions will make failure more of a certainty. As I have been known to say to my students, each time you accept an obverse action, you move a step closer to ensuring that you are moving down that long, lonesome road to failure.

10.1 THIRTY-SIX WAYS TO FAIL

To make this perspective more concrete, we look next at my list of three dozen ways to move one inexorably in the direction of failure:

Managing Complex Systems: Thinking Outside the Box, By Howard Eisner
Copyright © 2005 John Wiley & Sons, Inc.

1. Promise and market products you don't have.
2. Deliver products before they're ready.
3. Be everything to all people in the marketplace.
4. Approach government markets and commercial markets in exactly the same way.
5. Don't aspire to be the best in anything; be average in everything and then grow your lack of special competence into a big business.
6. Believe the saying: "Build the product and they will come."
7. Do most everything ad hoc.
8. Make sure not to learn from your mistakes.
9. Don't learn from other people's mistakes.
10. Make sure that you listen to outsiders more than insiders.
11. Make sure that engineering and marketing personnel never talk to each other.
12. Always make your technical decisions on political grounds.
13. Always overpromise and underdeliver.
14. Work with no cash flow margin so that you're always on the brink.
15. Keep your plans ambiguous and under wraps so that you can change direction and commitments at a moment's notice.
16. Reinvent the wheel as much as possible.
17. Penalize your people who complain, and declare them not to be team players.
18. Work your engineers 12 hours a day, six days a week.
19. Make sure to avoid focusing (ready, fire, aim).
20. Forget about building effective teams.
21. Don't ask your lead engineers for their opinions about key issues; assume they're too busy working to be interested in the "big picture."
22. Rely more on your staff people than on your line people.
23. Keep changing direction based on the last idea from some vice president.
24. Always assume that your people don't understand and don't really care about the company's best interests.
25. Conclude that your people are wrong any time they disagree with you.
26. Confirm that planning is antithetical to action, so don't do much of it.
27. Assume that communications within the company are about as good as they can be.
28. Make and then try to meet unrealistic schedules.
29. Never ask for multiple independent estimates of cost and schedule.
30. Always work schedules and budgets without contingencies and reserves.
31. Assume that the first functional decomposition of the system is correct and inviolate.

32. Assume that all the requirements provided by the customer are correct and inviolate.
33. Never do a formal risk assessment or mitigation analysis.
34. Never consider special incentives for delivery within budget and schedule.
35. Assume that your real managers don't ever need extrinsic motivation or recognition.
36. Turn your back on your customer after the sale is made or the system is delivered.

A close look at the first six items in this list shows that they tend to be related to business perspectives and marketing. The next 10 are basically process-oriented. The following 11 are concerned with work habits and team relationships. The next three deal with budgets and schedules. Statements related to systems engineering elements are in the next three. The following two deal with incentives and recognition. Finally, the last item relates to the customer. Rather than discussing every item in the list, in the following text we explore a selected top dozen ways to fail from the list of 36. Keep in mind the intent of this thinking perspective. Put a check mark next to those statements that appear to you to describe perspectives that are present, or possibly even dominant, in your organization. Also, try constructing your own list of at least a dozen statements that deserve to be added to the list.

10.2 TOP DOZEN OBVERSITIES

10.2.1 Do Most Everything Ad Hoc

The degree of ad hoc behavior tends to be inversely related to the size of an enterprise. Small companies tend to comply with this obversity, and depending on how small they are, it is hard to argue with such an approach. Quite large companies tend to have a procedure for almost everything, and that can obviously be taken to an extreme.

The real focus here is on *process*, a word that has received a great deal of attention in the past 15 years or so. It came alive in bright colors with the advent of *business process reengineering* [10.1], a set of activities considered quite valuable by both industry and government. The essential notion is that if you're not satisfied with the results you're getting, change the process that led to those results. It is difficult to argue with this concept, although there has been slippage between the concept and its implementation.

Another force behind following a process that works has been the various capability maturity models that have been developed and applied, especially by the Software Engineering Institute of the Carnegie Mellon University [10.2]. These models define process areas (or key process areas) that appear to be critical to success. Accepting this selection and premise, such processes need to be instituted and followed. Areas of application have included software engineering, systems engineering, integrated process and product development, and the acquisition of systems.

I support the notion that process is important unless it is overdone and taken to an extreme. So a company and its executives need to find a balance that works. That balance point is likely to shift as the company grows in size and complexity of operation. Clearly, complete ad hoc behavior can be a formula for disaster.

10.2.2 Make Sure to Not Learn from Mistakes

Looking at our list in Section 10.1, we see that it applies both to your mistakes and to the mistakes of others. This obversity, of course, flies in the face of Senge's declaration that all enterprises need to become learning organizations [10.3]. Some of his suggestions include focusing on the following five core areas (see also Chapter 3):

1. Building shared vision
2. Team learning
3. Mental models
4. Personal mastery
5. Systems thinking

Senge calls these five areas *component technologies* that innovate learning organizations. He also takes each of these disciplines and identifies subordinate levels for consideration:

1. The *essence* of the discipline
2. The *principles* that guide the discipline
3. The *practices* in terms of specific recommended actions

We do not, of course, have a good measure of the extent to which this obversity is or is not followed. We can assume that day-to-day decisions all around the world are tempered: "We don't want to repeat the mistake we made last year," or "The XYZ folks tried that and went down in flames, so let's not go down that road." In other words, it would appear that just about everyone will try to avoid mistakes, but that requires a clear identification of some particular action as a mistake. Look at the following list of a dozen "mistakes" and see if you agree that they all were, indeed, mistakes.

1. Ford's Edsel automobile
2. Wang Labs' failure to recognize the potential impact of open hardware and software systems for the personal computer
3. Time-Warner's merger with AOL
4. Hewlett-Packard's purchase of Compaq
5. IBM's entry into the personal computer market

6. Failure of the railroads to go into the air transportation market
7. Kendall's (Digital Research) apparent lack of interest in supplying an operating system for IBM's personal computer
8. Dan Bricklin's apparent disinterest in cashing in on his invention, the Visicalc spreadsheet
9. General Motors' purchase of Electronic Data Systems (EDS)
10. The venture capital industry's financial support during the Internet bubble
11. IBM agreeing that Microsoft share in the proceeds of PC-DOS, which they sponsored
12. The sale of McDonnell Douglas to Boeing

Another factor that is related to our propensity to learn from mistakes has to do with the mental models that we develop and carry around with us. This can be the same mental model that Senge talks about, or it can be the type of model that is based on one's previous experience, which might neglect current facts and possible likelihoods [10.4]. Neglecting or pushing immediate facts to the background is a dangerous action that will have a tendency to support this particular obversity.

10.2.3 Always Overpromise and Underdeliver

Many companies appear to overpromise to their customers, and wind up under-delivering. They do this as a business practice, operating under the assumption that doing so will maximize the chances of a sale. To do otherwise, they claim, makes them vulnerable to their competition.

The footnote to this particular obversity is that it came originally from a question that I asked my son. He had been achieving considerable success in a well-known high-tech company, building new systems for them. I asked him if he could cite just the one thing that contributed most to his success. After a minute or so, his answer was: "Always underpromise and overdeliver." Now his answer is appearing here as an obversity. Apparently, it also served him well in the real world.

By overpromising, of course, we often place extreme demands on the company of which we are a part. An illustrative scenario might play out in this fashion. A salesman for XYZ Electronics is visiting with a customer. The customer asks about a feature of an electronics black box that is not listed in the product spec sheet. The salesman declares that whereas the feature is not currently in the box, it can be added and delivered within three months if the customer order is sufficiently high. The customer says: "If you can deliver that feature within three months, I'll buy 300 boxes at the price we discussed." "You've got a deal," says the salesman.

To make good on what he knows to be a serious overpromise, the salesman comes back to company headquarters and tries to convince the powers that be that the company must respond to this wonderful opportunity. Lots of new revenues, a new customer, a new feature in the product, and a commission to the salesman. All of this could be true, except for one minor detail. The company has about a 1% chance of being able to respond. And if it were somehow to do so, lots of people would be working overtime for months. The scenario has many variations illustrating several outcomes, most of which lead to underdelivery and an unhappy customer.

Part of the notion of rejecting this obversity is to work seriously at truly determining when a product can be delivered, at what cost, and with a relatively high likelihood of success. Then a quite reasonable promise can be made, which may well lead to a win–win solution. By promising less, we are hopefully managing the customer's expectations. This topic is itself worthy of an entire chapter in a primer on marketing and sales.

10.2.4   Reinvent the Wheel As Much As Possible

This obversity can take several forms. One is simply that of the development engineer who loves to design new devices but hates to do the searching necessary to find out if the device already exists, in whole or in part. It's more fun to reinvent the wheel, especially since the company won't mind if it happens that way. This issue was explored in Chapter 6 with respect to software reuse. There are certainly many reasons why building new software from scratch can be more fun than working the reuse scenario. But is "fun" really the objective or the goal? I think not.

Some companies reinvent the wheel not knowing that they are doing so. To the extent that this is the case, all tasks become new challenges and the leverage achieved is minimal. The matter of creating leverage was discussed in Chapter 6. One of the key responsibilities of management is to assure that the enterprise achieves increasing amounts of leverage, consistent with the general and ethical principles of building a business.

The matter of reinventing the wheel is also embedded in the field known as *knowledge management* (KM) [10.5]. Some of the more serious aspects of KM is for organizations:

1. To know about the overall knowledge base that is "owned" by the organization
2. To have access to that knowledge base
3. To make decisions as to when and where to create or acquire new knowledge

If and when the organization gets to item 3, it is able to decide *not* to reinvent what has already been invented. However, we may presume that many organizations get to item 3 without a good handle on the first two items. If so, they may decide, with an incomplete understanding of the situation, to come to a less than optimal conclusion.

10.2.5   Make Sure to Avoid Focusing

There are times when the "buckshot" approach is better than the "rifle-shot" approach, but not many of them. Focusing will usually lead to better results in the long term. This statement can be supported by a very well known and widely practiced activity: strategic planning. Here are some recommended elements of strategic planning from George Steiner, one of our leading experts [10.6]:

1. Strengths
2. Weaknesses
3. Opportunities
4. Threats

This has come to be known as *swot analysis*, for obvious reasons. The purpose of these rather specific forms of analysis is to provide as much focus as possible in areas that are critically important to the enterprise.

On the other hand, many enterprises prefer an approach that fosters creativity and innovation: one that allows everyone to do what they wish to do. But we tried that with the Tower of Babel and it didn't seem to work. Those who like this particular obversity can join with like-minded folks who follow this approach with a simple mantra: Ready ... Fire ... Aim.

10.2.6 Confirm that Planning Is Antithetical to Action, So Don't Do Much of It

Although good planning helps an enterprise to establish and maintain focus, some people believe that planning is almost always overdone and that it creates a mindset that prevents taking action in the real world. Certainly, there must be companies and individuals who use overplanning as a way to avoid action, but we must assume that these are exceptions whose example is not to be copied.

The well-functioning organization has planning at both the strategic and tactical levels. A brief citation of Steiner's substantial contribution to the former was dealt with in Section 10.2.5. My analysis of strategic planning in a university environment was documented [10.7] and led to the following recommended steps in the process:

1. Start with bottom-up planning at the department level.
2. Confirm with top-down guidance.
3. Expand dialogue to school-wide forums.
4. Provide team training and facilitation.
5. Allow time for threefold iteration.
6. Provide feedback and communication.
7. Document.
8. Emphasize measurable goals.
9. Confirm commitment to goals.
10. Look for and support champions, leaders, and team builders.
11. Assure incentives and recognition.

12. Handle the dissidents.
13. Negotiate and confirm budget–performance coupling.
14. Measure results each quarter.
15. Sustain yearly continuous improvements.

This was followed by the articulation of sections of the document known as the *strategic plan*. Of particular note was the setting of goals in such areas as:

1. Education
2. Research
3. Publications
4. Enrollments
5. Personnel and student development
6. Infrastructure and organization
7. Productivity
8. University/community service
9. Incentives and recognition

The elements of a tactical plan are somewhat more difficult to set forth since tactics depend largely on the local situation. However, there is an organization that has studied this issue for a very long time and has proven over the years that it not only knows how to plan in a tactical sense, but also knows how to implement such plans. That organization is the Department of Defense.

10.2.7 Make and Then Try to Meet Unrealistic Schedules

One of the most prevalent complaints that one sees, especially in the high-tech world, is that everyone is working long hours in an attempt to meet impossible schedules. For some reason, this mode of behavior appears to be standard operating procedure on software development projects. This obversity calls for continuing this pattern until everyone is frustrated, burned out, or both.

Impossible schedules can come about from at least the following two rational sources:

1. The customer demands results by a certain date.
2. Management believes that they are necessary in order to beat the competition.

These are called *rational* because they can possibly be justified on logical grounds to those who are working overtime on a continuing basis. The setting of impossible schedules that appear to be arbitrary and capricious to the software engineers will often create lots of bad-will as well as loss of valuable employees.

The systems and software communities have worked hard to establish bases for rational schedule and cost estimation. To illustrate, the software development practitioners can rely on two well-known methods for estimating schedule and cost:

1. COCOMO (constructive cost model)
2. Function point analysis

Proponents of these procedures are continuously looking for real-world data that will help to improve the quantitative relationships that support the estimation process. This helps to keep the processes current, correct, and rational. To illustrate further, assume that a software development project that is to deliver 35,000 source instructions needs to be scheduled. Under a reasonable set of assumptions using COCOMO, estimates of person-months and development time are forthcoming (see also Chapter 8) as [10.8]

$$PM(\text{person-months}) = 2.4(KDSI)^{1.05} = 2.4(35)^{1.05} = 100.3 \text{ person-months}$$
$$\text{Development time} = 2.5(PM)^{0.38} = 2.5(100.3)^{0.38} = 14.4 \text{ months}$$

Schedules that significantly compress the recommended development time of almost $14\frac{1}{2}$ months are likely to be unrealistic, creating negative effects and consequences within the enterprise. When we have tools that will help us make better programmatic decisions, it behooves us to use them in our attempt to behave as rationally as possible and obtain the best possible results.

10.2.8 Never Ask for Multiple Independent Estimates of Cost and Schedule

The early COCOMO equations cited above show that it is possible to estimate cost and schedule from an input of delivered source instructions (DSIs). The two important outputs depend on one critical independent variable input. In this and similar situations, we have to keep in mind the old and well-known wisdom: garbage in, garbage out (GIGO). If the DSI input is far from what turns out to be the true value, the cost and schedule estimates will be off similarly. This is a most important fact that needs to be considered when using COCOMO or any other model to estimate cost and schedule.

The point then becomes one of how to obtain the best possible input estimate of DSI in the example above. One very simple but effective approach is *always* to obtain multiple independent input estimates. A place to start is to request inputs from three different people, asking each to prepare the estimate without talking to the other two estimators. Each is provided with the same information about the project, including the requirements specification and all documents provided by the customer. Give the estimators the same time to respond, such as a few days. Each should be prepared to back up the estimate with whatever material each finds necessary. In other words, each needs to formulate a specific rationale for the input that each has selected.

The inputs should be revealed at a meeting of all parties. All estimates are presented, together with the reasons. This usually leads to rather detailed discussions of the points that they all agree with, as well as points of contention. Each point of disagreement should be discussed in considerable detail, so as to reach as much agreement as possible. One is looking for a consensus answer that all can subscribe to in the sense that all could see how the project could be completed with the consensus estimate. If such a result is not workable, another two to three estimators should be brought into the process. You are looking for a result that works, with sufficient independence of view so that risk factors and similar considerations are taken into account. Failure to be very careful about selecting the appropriate inputs could well lead to many regrets down the road. A lockup on a bad number affects cost and schedule estimates in a clearly negative way. This is not a good way to start off a new project.

10.2.9 Assume That the First Functional Decomposition of the System Is Correct and Inviolate

Functional decomposition is an early step in the design of a complex system. It identifies the key functions and subfunctions that form the basis for system architecting. Figure 10.1 illustrates the functional decomposition of a simple system, the desktop computer.

The precise method of architecting is explored in some detail in Chapter 13. Suffice it to say at this point that functional decomposition and architecting are critically intertwined. If we accept this as a premise, it is fair to conclude that *if the functional decomposition is incorrect, the system architecture is likely to be incorrect.*

Proceeding with a flawed architecture will usually lead to problems during the development and testing of the system. All of this implies that the functional decomposition of a system should be under scrutiny as long as possible in an attempt to avoid locking up on a decomposition that may be incorrect. This is

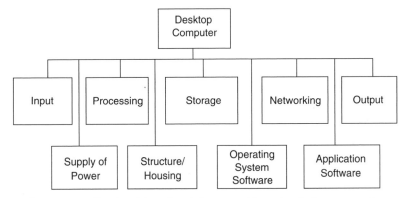

Figure 10.1 Top-level functional decomposition of a desktop computer.

related to one of Rechtin's heuristics [10.9]: *Build in and maintain options as long as possible as you design and implement complex systems.*

It has also been argued that the functional decomposition of software systems is less obvious than it is with hardware systems. If this is true (and I tend to agree), even more care needs to be exercised in the functional decomposition of a software system. Indeed, the decomposition is a serious part of the design and architecting process. Excellent guidelines that we have been given are from Fred Brooks [10.10] and Niklaus Wirth [10.11], who suggest, respectively, that:

- From Wirth's top-down methods, one identifies *modules* of solutions or of data that can proceed independent of other work [10.10].
- The most difficult design task is decomposition of the whole into a module hierarchy [10.11].

We are interested in decomposing functions such that there is minimal interaction between parallel functions. This allows teams to work on these parallel functions as independent of one another as possible. Independence simplifies the design as well as its implementation and testing. The bottom line is simply that a correct functional decomposition is critical, and changes that need to be made in a decomposition are the price that we should be prepared to pay in order to "get it right."

A further example of some of the vagaries of functional decomposition can be seen by a brief look at the matter of designing a project management (PM) software system. With even a cursory knowledge of PM functions, we may be led to the functional decomposition shown in Figure 10.2a. This figure emphasizes the

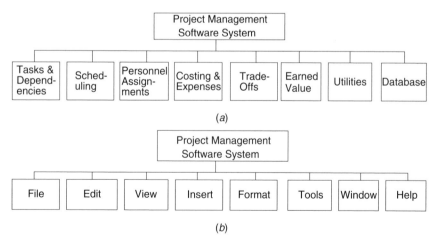

Figure 10.2 Top-level functional decompositions of project management software: (a) generic project management functions; (b) as implied by Microsoft project software systems. (From [10.12].)

following as key top-level functions:

1. Tasks and dependencies
2. Scheduling
3. Personnel assignments
4. Costing and expenditures
5. Trades and optimization
6. Earned-value analysis
7. Utilities
8. Database

From another perspective, we can also look at the top-level functions implied by the Microsoft project menu items [10.12]. The latter are also functions and are illustrated in Figure 10.2b. The functions known as *file, edit,...* and so on, are well known to users of various types of software products, and it is not unreasonable to use them as a basis for the decomposition of this software. We note the differences between the approaches shown in the two parts of Figure 10.2.

10.2.10 Assume That All the Requirements Provided by the Customer Are Correct and Inviolate

The issue with respect to requirements in many ways parallels the functional decomposition matter discussed above. Some people indeed believe that the initial set of requirements should be accepted as correct and inviolate. On that basis, requirements creep and volatility are pointed to as reasons why a development program has gone astray. Another way to look at this issue might be simply to recognize that we know the least about the requirements of a complex system at the beginning of a program. Since we must have a place to start when defining a system, requirements is clearly a logical point of departure. However, as we progress further into a program we learn more and more about the system. As we become "smarter," it makes sense that we should be able to challenge and modify poor requirements rather than consider them unchangeable. What is needed is a flexible set of processes that accepts this necessity without labeling it as a reason for failing to build systems correctly. As with the functional decomposition, if we set a bad requirement in concrete, we are likely to have to pay the price somewhere down the road. Which path would you most like to select: one that creates problems by failing to acknowledge and change bad requirements, or one that creates problems by modifying bad requirements as soon as they are determined to be bad?

This obversity of assuming that initial requirements are correct and inviolate, it is suggested here, will lead to no end of trouble. We all try to start a program with a perfect articulation of requirements. Empirical data appear to indicate that this is the exception rather than the rule. Some government practices, such as spiral development, evolutionary acquisition, and capabilities-based development, all support the notion that requirements should be subject to change. Some of the systems that we have in place (e.g., procurement practices) are obstacles to doing what we need to do.

10.2.11 Never Do a Formal Risk Assessment or Mitigation Analysis

Although much has been written about risk analysis and mitigation and their importance, it appears that many system development programs get up and running with no formal attention to matters of risk. An example of this lack of attention is revealed by a brief examination of the first software capability maturity model (CMM) [10.2]. This model cited 18 key process areas (KPAs), shown in this book in Section 6.5.1, which by definition represented areas of the most serious focus when developing software. The list of KPAs does not include risk analysis and mitigation. It is heartening, however, to recognize that this oversight was repaired by the time the integrated CMM model [10.2] was developed.

As systems become more complex, evolving into families and systems of systems, it would appear that formal risk analyses will become increasingly important. For many of our systems that place human lives in jeopardy, the necessity of risk analysis and mitigation is emphasized. Even so, we do not have to go further than the two NASA missions (*Challenger* and *Columbia*) in which failures led to the loss of entire crews. Surely, it is time to truly adopt a philosophy that implements the old saw "an ounce of prevention is worth a pound of cure." All developers of new and complex systems should have to watch the videos of those two NASA disasters. It's time to slow things down a bit and apply the considerable engineering talent that we have to truly mitigate the risks of many of our systems and the potential jeopardy of their related missions.

10.2.12 Assume That Your Real Managers Don't Ever Need Extrinsic Motivation or Recognition

The complexity of our new systems has increased the level of difficulty associated with engineering these systems. Much like the concept of degree of difficulty with respect to certain Olympic events (e.g., diving), our best system builders are proficient at tasks that are themselves quite difficult. These people are normally in great demand in an enterprise, as they can literally make the difference between success and failure on one or more projects. They are often called in at the beginning to set a project on the right course or to solve a particularly knotty problem. At times, they are asked to come onto a project that is in serious trouble, the agenda being to find a "bail-out" solution.

In today's often homogenized world, there are times when we ask our best people to perform yet another miracle on behalf of the company, without special consideration for service beyond the call of duty. This particular brand of obversity will usually and ultimately lead to two negative consequences:

1. Our best people refuse the privilege of converting to a six-day, 10-hour work-week.
2. Our best people find employment elsewhere.

Special people need to be treated in special ways, and most enterprises are able to respond to that need. If no response is forthcoming, it is usually a result of just not thinking about it, or deciding that a special response is not called for. Successful enterprises make appropriate use of such mechanisms as bonuses, out-of-cycle raises, stock options, and other forms of recognition. Not rewarding special performance in one way or another reinforces this obversity and leads an enterprise down the wrong road.

10.3   SUMMARY: A MEETING

John, a middle manager at the M&O Software Development Company, remembered that the last monthly measurements meeting at which the status of his project SP1 was reviewed in detail by his vice president. The results were not good. The project was two months behind in schedule and 6% over budget. Steve, his project manager for the project, seemed a bit shaky, but he did not want to manage Steve's day-to-day steps. But the next monthly review was only four days away and John decided to check in with Steve to make sure that things were getting better. He scheduled a meeting with Steve that afternoon.

John: So how're we doing on SP1?

Steve: We're seeing a lot of daylight, I'm happy to report.

John: Do you have numbers for the measurements meeting?

Steve: The numbers I'm getting say that we're about three months behind and about 11% over budget. I know that's worse than last month, but this type of turnaround takes time.

John: Frankly, Steve, those kinds of numbers are shocking. How can you explain them?

Steve: Well, I've taken three major steps to solve the problem. First, I brought three new people onto the project to help shrink the schedule. Second, I've put the entire team on 10-hour workdays, with a half day on Saturday. Everybody's writing code with both hands. Finally, I've increased the scope of the project somewhat so that we can show the customer more capability, which will justify the extra monies we've spent.

John: Steve, these actions make me even more nervous. Have you filed a project risk report, according to our standard procedure in cases like this? I've not seen one.

Steve: Risk? We're in a minimal-risk situation. No need for such a report.

John: Well, how is it that the project status has become worse?

Steve: Can't tell exactly, except that I didn't really expect better news for awhile.

John: Have you taken another look at the system breakdown structure and decomposition?

Steve: Too late for that. We're moving aggressively with what we've got.

John: Well, what do you forecast for the end of next month, a month after the meeting in four days?

Steve: Don't know the answer yet. Can't seem to get any usable data from my lead software engineer.

John: I'm not happy about where we are. I think we need a lot better answers for our measurement meeting next Tuesday.

Steve: I'm running about as fast as I know how and pushing the troops as hard as I can.

John: I'm afraid that what you've been doing isn't solving the problem. I'm going to call an all-hands meeting with your team first thing in the morning. I'll chair the meeting; you let everyone know by e-mail.

Steve: I thought we agreed at the beginning that you weren't going to micromanage my project.

John: We did, and I'm not. As of close of business today, it's no longer your project, it's mine.

Too much obversity and several new steps in the direction of failure were enough to change horses.

REFERENCES

10.1   Hammer, M., and J. Champy (1993). *Reengineering the Corporation.* New York: HarperBusiness.

10.2   See Web site at sei.cmu.edu.

10.3   Senge, P. (1990). *The Fifth Discipline.* New York: Doubleday Currency.

10.4   Kahneman, D., P. Slovic, and A. Tversky (eds.) (1982). *Judgment Under Uncertainty: Heuristics and Biases.* Cambridge: Cambridge University Press.

10.5   Harvard Business Review (1987). *Harvard Business Review on Knowledge Management.* Boston: Harvard Business Review Paperback.

10.6   Steiner, G. (1979). *Strategic Planning,* New York: Free Press Paperbacks.

10.7   Eisner, H. (1998). University Strategic Planning: Recommended Product and Process, presented at the American Society for Engineering Management's 19th Annual Conference, Virginia Beach, VA, October 1–3.

10.8   Eisner, H. (2002). *Essentials of Project and Systems Engineering Management*, 2nd ed. New York: Wiley.

10.9   Rechtin, E. (1991). *Systems Architecting.* Englewood Cliffs, NJ: Prentice Hall.

10.10   Brooks, F., Jr. (1995). *The Mythical Man-Month.* Reading, MA: Addison-Wesley Longman.

10.11   Wirth, N. (1995). A Plan for Lean Software, *IEEE Computer Magazine*, February, pp. 64–68.

10.12   Microsoft Project (1994). *Operating Manual*, Redmond, WA: Microsoft.

Chapter 11

Perspective 7: Remove Constraints

Many problems cannot be solved as a consequence of too many constraints. Similarly, many systems cannot be constructed and managed properly as a result of the imposition of too many constraints. This particular way of thinking outside the box formally acknowledges the constraint issue and addresses it directly.

Examples of inappropriate constraints that we have seen in our own or other people's lives include judgments about them such as:

- He's too old to get the job done.
- Her handicaps will keep her from delivering for us.

With reference to the former, I cite the counterexamples of Bertrand Russell and Colonel Sanders. Helen Keller clearly demonstrates that even very severe handicaps can be overcome. So does Stephen Hawking, the physicist and professor from the U.K.

In the world of large-scale complex systems, constraints come in several forms, most often associated with schedule, cost, and performance, from a top-level perspective. Accepting undue or inappropriate constraints keep us inside the box. Challenging and removing such constraints give us part of the wherewithal to move outside the box.

Managing Complex Systems: Thinking Outside the Box, By Howard Eisner
Copyright © 2005 John Wiley & Sons, Inc.

11.1   TYPICAL SYSTEM CONSTRAINTS

Some dozen typical constraints that might be experienced when dealing with systems are the following:

1. Not enough funding
2. Insufficient time
3. Poor driving requirements
4. Insufficient technical expertise
5. Poor facilities
6. People not team players
7. Insufficient financial information
8. No supporting tools
9. Poor subcontractors
10. Negative attitudes
11. Counterproductive customers

Each of these areas is explored briefly below.

11.1.1   Not Enough Funding

Insufficient funding is certainly one of the "longest poles in the tent," and it is an issue that has several dimensions. Removing this constraint might simply mean getting additional funding—usually more easily said than done. Another way to remove the constraint is to figure out how to get the job done at the current level of funding. Yet a third approach is to scale the job back (i.e., reduce system scope) so that the allocated funding is sufficient.

Obtain Additional Funding   Whatever the source of funding, and assuming that additional funds are in fact available, the sponsor needs to be convinced that job (e.g., system) completion is in serious jeopardy without the allocation of additional money. This usually means a lot of homework to demonstrate clearly that the tasks at hand have indeed been underfunded. Independent cost estimation through cost-estimating relationships can be helpful in this regard. So can new empirical data that support the estimates.

Reevaluation So That Funds Are Sufficient   This basically involves removing the constraint by finding better ways of getting the job done. It usually requires very detailed reevaluation of all work tasks, looking for ways to save money without changing scope. Perhaps you can pay less by buying in quantity. Possibly a lower-cost supplier can be located. Maybe a more elegant approach and solution can still be found, one that saves money while also maintaining performance. We note that many companies appear to have been driven to have software created overseas to save money over the cost of a more conventional approach.

Scaling the Task Back   This approach is usually not acceptable without the concurrence of the customer or sponsor. As with the first idea above, homework is necessary to show that there is no conceivable way to (1) get the job done within the current job scope, and (2) define a "plan B" that would result in a scaled-back effort that would give the customer a highly desirable output.

11.1.2   Insufficient Time

Although insufficient time and not enough funding can "travel together," consider now only the case of the former as a constraint. As might be expected, the "solutions" that remove this constraint parallel those expressed above under insufficient funding:

1. Have the customer agree to an expanded schedule that is workable.
2. Compress to an acceptable schedule by means of a more efficient work plan.
3. Reduce the scope of work so that the current schedule can be met.

For the first of these, the constraint has gone away completely. The second often suggests doing more work in parallel, which usually requires adding more people who can be highly productive. In this regard, one must consider Brooks's law [11.1] for software, which says that adding more people to a project that is late will often make the project even later. Finally, reducing the scope may turn out to be the last resort, and many project managers and customers wind up with this as the only practical solution.

It would appear that no small number of system efforts are forced to sign up to an artificially derived schedule that bears little relationship to the realities of what can actually be accomplished under the given circumstances. Then schedule slippage begins to occur and there is much consternation as well as questions about how this situation came about. Really bad and unrealistic schedules can be serious de-motivators for a project team and are clearly not recommended. Tough schedules can be stretched for, but only if (1) most everyone knows and basically agrees with the reason for such a schedule, and (2) bonuses are available for the key players who make it happen. Working 12 hours a day to meet a schedule that is viewed as arbitrary and capricious grows old in a hurry and can create considerable damage to an otherwise highly productive team.

11.1.3   Poor Driving Requirements

Driving requirements for a system are those requirements that have an extremely large impact on the system architecture and its implementation. For example, a system requirement for a 98% availability is usually not a key driver. When pushed an additional two orders of magnitude, the system can be down only 0.02% of the time, and this would normally convert into a key driver. Considerable standby redundancy must be added to meet such a requirement, as downtimes are truly minimized. Driving requirements can also be thought of as related to:

1. A substantial increase or reduction in system performance.
2. The forward edge of the state of the art.
3. Increasing risk, as a consequence of requirement 2.

Removing this type of constraint means that it should be possible to modify driving requirements significantly, as appropriate, to increase the likelihood of program and system success. As noted in Chapter 1, inside the box thinking treats requirements as fixed and inviolate. Thinking outside the box in this regard means that requirements can be changed when it is in the best interest of the overall system design and construction to do so. This notion is also in consonance with more recent views of the evolutionary development of systems [11.2].

One very specific example of removing a constraining requirement was discussed in some detail in Chapter 1. The requirement was to construct a "best of breed" system, with the following subsystems:

1. A database management system (DBMS)
2. A word processor
3. A spreadsheet
4. A Presentation manager

For this example, a study had been carried out that identified the best of breed for each of the above, yielding the following results:

- DBMS: Oracle
- Word processor: Wordperfect
- Spreadsheet: Lotus 1-2-3
- Presentation manager: Microsoft's Powerpoint

Moving forward with the best-of-breed concept, would therefore mean integrating these four systems, each of which is produced by a different company. This is clearly a most difficult and prodigious task. However, if we remove the constraint that we must integrate the systems found to be the best of breed, we are free to look at Microsoft Office or Lotus's Smartsuite as candidate systems that have the required four-subsystem functionality. Removing the constraint in this case brings us from a poor design concept and almost certain failure to a system solution that is more than satisfactory.

Another illustration of removing a constraint can be drawn from the history of Microsoft and can be seen as one of its defining moments. In 1979, IBM came to Microsoft to sign them up to produce an operating system for their PC [11.3]. Since Microsoft considered themselves in the applications business rather than in the operating system business, they sent IBM off to Digital Research, who had built the CPM operating system. When the meeting between IBM and Digital Research was unsuccessful, IBM went back to Microsoft. At that point, Bill Gates and Steve Ballmer removed the constraint that said they were not in the business of building operating systems. This resulted in a contract between the two companies that ultimately led to PC DOS and MS DOS. This pivotal moment in the life of Microsoft paved the way for their ultimate success with operating systems and Windows. It can be said that all of this success was based on knowing when to remove a constraint, one that turned out to be artificial and no longer useful. Thinking outside the box in this particular way at that time led to a multibillion-dollar decision and the pathway to a singular success story for Microsoft.

11.1.4   Insufficient Technical Expertise

Not having the appropriate technical expertise to build and manage a system will almost certainly lead to failure. How this constraint may come about is probably not too important. What is critical, however, is that it cannot be accepted. This is a "push-to-the-wall" situation: a do-or-die issue for a project manager (PM). Here are some of the possible consequences of not having the required technical skills:

1. Inability to question requirements appropriately
2. Poor system architectures
3. Poor subsystem design
4. Poor record in head-to-head competitions
5. Inability to satisfy technical milestones
6. Slippage in cost and schedule, as a result of item 5

If we are talking about a project or equivalent, the PM must have the authority to select technically competent people to get the job done. Conversely, less-than-competent people cannot be foisted on the PM. That is, he or she must be able to reject such people and have hiring and firing authority. Thus, the perhaps evident ways to remove this possible constraint is to assure such authority and to confirm the PM's ability to know the differences between various levels of technical capability.

11.1.5   Insufficient Facilities

Insufficient facilities can range from being a minor inconvenience to being a critical obstacle. In the former category, for example, is the temporary relocation of a project team from one facility to another. Even if the new facility is more spacious and usable, some members of the team will find it an inconvenience, if for no other reason than that they must travel longer to get to work. At the other end of the spectrum, for example, is the lack of a vacuum thermal chamber in which to conduct tests of a satellite, or not having a shake table to use for required vibration testing. Clearly, the constraint represented by the latter cases must be overcome so that the full scope of the test programs can be addressed. Failure to do so would compromise the integrity of the satellite system and the confirmation of its performance during launch and in space.

Removal of such significant constraints can be overcome relatively easily by (1) leasing the required facilities from a third party, or (2) bringing such a third party onto the team during the bid process. Thinking inside the box says that there is no way to make a credible bid on the satellite contract. Moving outside the box removes the constraint by engaging a third party and making arrangements to use its facilities for testing. Persistent removal of constraints tends to make available courses of action that were previously thought not to be possible.

11.1.6 People Not Team Players

The development and management of complex systems in today's world is now highly dependent on the use of teams. This is supported by the relatively large number of books on the topic of building teams, which go by several related names:

1. *Executive Teams* [11.4]
2. *Empowered Teams* [11.5]
3. *The Wisdom of Teams* [11.6]
4. *The Team Handbook* [11.7]
5. *Effective Team Building* [11.8]

A considerable recent focus is on the matter of *integrated product and process teams* and *high-productivity teams.* These involve a high level of teamwork by which people work well together, help each other, and achieve extraordinary productivity. One or more members of your team who are not team players may be able to undermine the entire team. Such "team busters" represent a possible severe constraint for the program or project manager to contend with. The bottom line is that removing team busters is mandatory to set the stage for success. This means, of course, that the PM must have the authority to do so, but may have failed to build a case that clearly demonstrates "team buster" behavior. This type of behavior is well known [11.9], although the team buster is often very clever and convincing in his or her negative representations.

11.1.7 Insufficient Financial Information

Trying to run a project, program, system, or enterprise in a successful manner normally requires significant amounts of financial information. Examples of such information follows:

1. Total cost expended, last time period and cumulative
2. Total direct labor costs expended, last time period and cumulative
3. Other direct costs expended, last time period and cumulative

4. Overhead costs
5. General and administrative costs
6. Foregoing costs by task
7. Foregoing costs by work breakdown structure element
8. Foregoing costs by labor category
9. Foregoing costs by calendar and fiscal year
10. Foregoing costs for each project, by program

At the corporate level, it is often the case that *both the CFO and the CIO* need to work together to provide timely and accurate information since both typically report to the CEO (or the COO). They are likely to give very high priority to those top-level needs. Partly as a consequence, the program or project manager can experience a lack of financial data, which can represent a most serious impediment to success. Furthermore, since such information is provided from a separate chain of command, many inexperienced managers are at a loss as to how to remove this type of constraint.

At the project level, the outside-the-box answer, perhaps unfortunately, can be to capture certain types of information at the source and use one or more spreadsheets to process and display the results. A good example is the PM who is trying to obtain a weekly report on direct labor cost expenditures. The time card hourly data can be collected every Friday afternoon, and then fed into spreadsheets that convert hours worked into cost expended, by person, by personnel category, and by task area or work breakdown structure element. All of this information can literally be in the hands of the PM by a half hour after close of business, with graphical outputs. This "homemade" tracking system is a way to remove the apparent constraint of poor or untimely data that might be provided by the company's project cost information system.

Another aspect of providing financial information is related to the notion of earned-value analysis [11.9]. This approach focuses on three key measures of cost:

1. Budgeted cost of work performed (BCWP)
2. Budgeted cost of work scheduled (BCWS)
3. Actual cost of work performed (ACWP)

On this basis, the cost variance (CV) and the schedule variance (SV) are calculated as

$$CV = BCWP - ACWP$$
$$SV = BCWP - BCWS$$

This short discussion is a brief example of the need for certain types of financial information in order to be successful. Constraints that prevent the collection, analysis, and use of these types of information need to be removed systematically.

11.1.8 No Supporting Tools

A constraint that consists of the lack of supporting tools has an excellent chance of leading to failure. In today's world of large-scale systems, many supporting tools take the form of software that facilitates both systems and software engineering. Some of the systems and software engineering tasks that are regularly supported by such software tools are as follows:

1. Requirements engineering
2. Technical performance measurement
3. Modeling and simulation
4. Reliability–maintainability–availability
5. Life-cycle cost estimation
6. Configuration management
7. Specialty engineering
8. Software engineering
9. Software design
10. Software coding
11. Complexity measurement
12. Risk analysis

Having the right tools available to assist both systems and software engineering is considered a necessity in today's world of complex systems. If this represents a constraint in your company, you might consider drawing up an immediate investment plan to acquire sets of tools that apply most directly to your business enterprise. As new funds become available, they can be utilized to invest in tools in whatever sequence seems to be most cost-effective. Failure to remove this constraint may well leave your enterprise in the dark ages compared with your competitors.

11.1.9 Poor Subcontractors

As systems are becoming larger and more complex, systems integrators are forming large teams to tackle the matter of architecting and building such systems. This means that the prime contractors tend to have more subcontractors than ever before. There is no evidence that this trend is likely to reverse any time soon. More attention therefore needs to be placed on (1) selecting superior subcontractors to be part of your team, and (2) assuring that your subcontractor teammates are getting the job done.

The importance of subcontractor management is also reinforced by its explicit inclusion in the capability maturity model (CMM). The CMM for software-listed subcontractor management as one of its 18 key process areas [11.10] applicable at level 2 (see Chapter 10). The integrated capability maturity model (CMMI) includes supplier agreement management as one of its process areas at level 2.

11.1.10   Negative Attitudes

A typical system and business scenario is that an employee has a new and exciting idea to which the boss reacts negatively. This type of reaction will, of course, have predictably negative effects on the employee. Such effects can range from never again offering any good ideas, to asking for an internal transfer, to leaving the enterprise.

Employees with good ideas experience a boss who has consistently negative attitudes as a serious constraint, and the constraint can come in several dimensions:

1. The boss is a persistent obstacle to be overcome.
2. To move forward, a strategy is selected of "going around the boss."
3. In reacting to item 2, the boss becomes more negative, resulting in a downward spiral of negativism.
4. The employee holds back from suggesting new and better solutions.
5. As a consequence of item 4, the enterprise loses the benefit of these better solutions.
6. The enterprise experiences poor performance, resulting partially from item 5.
7. The employee is not promoted, partially as a result of item 4.

As one can see, negative attitudes can have very serious negative consequences if they are allowed to persist. Propagation throughout an enterprise represents a constraint. That requires positive mitigating action. The management team that fails to remove this constraint is headed for serious difficulties.

11.1.11   Counterproductive Customers

Many builders of new systems believe that the customer is always right. Although the customer holds a unique position with respect to the design and development of new systems, slavishly following the principle that the customer is always right can be a recipe for disaster.

There are, of course, times when the customer is right and times when he or she is wrong. When the customer is right, the development team must acknowledge this as the case and do whatever is required to support that position. The difficulty comes when the customer is wrong. A few such instances are listed below:

1. Not recognizing startup and learning curves
2. Insisting upon inappropriate requirements
3. Requesting impossible schedules
4. Trying to micromanage the contractor
5. Forcing the contractor to deliver a product or service at or below cost (no profit)
6. Requesting the delivery of work that is outside the scope of the contract

7. Using abusive behavior or language
8. Interfering unduly with the authority or autonomy of the contractor

In view of the fact that the customer can be wrong at times, companies need to be prepared to address such situations in as constructive a manner as possible. This can be viewed simply as removing the constraints that can result from a poor customer. It should also be remembered that sophisticated companies are also prepared to walk away from a customer when it is prudent to do so.

11.2 INTERNAL AND EXTERNAL CONSTRAINTS

As they approach the construction of large, complex systems, enterprises experience both internal and external constraints. Internal constraints often become reasons, right or wrong, why companies do not step up to a new challenge. External constraints may have the same effect, but they are based on factors outside the company. An example is the assessment that leads to a "no bid" decision when competition is viewed as being too strong. A related internal constraint is the possibility that internal capabilities to get a job done may be viewed as inferior to those of one or more competitors. This can translate into "Let's not waste money writing a losing proposal."

Enterprises need to watch out particularly for internal constraints, real or perceived, that will keep them from being successful. Approaches that see only problems, rather than opportunities, need to be challenged. Positive and "can do" attitudes should be nurtured. It is a great pity when enterprises fail to be "all that they can be" as a consequence of undue internal constraints. After all, it's the internal conditions that companies can do something about.

11.3 SUMMARY: A MEETING

For this particular approach to thinking outside the box, we postulate a meeting of four key executives in a large systems integration company that is doing a lot of contracting with NASA. These executives are the CEO, COO, CTO, and CIO. The CEO starts the meeting with the pivotal question that he would like to resolve.

CEO: So we are a key player in NASA's agenda. Now we need to come to terms with what the administrator is telling us. He wants us to build systems "faster, cheaper, and better" (F, C, and B).

CTO: He is clearly thinking outside the box. He wants contractors to find system solutions that make improvements in all three critical dimensions—simultaneously!

COO: I believe that means that unless we can do that, we'll wind up getting smaller and smaller shares of NASA's business.

CIO: It certainly sounds that way. And clearly that's not where we want to go. But is the F, C, and B approach feasible?

CTO: My people tell me, and I believe it, that two out of three may be workable, but three out of three basically isn't.

CEO: So what does that mean for us? Can we just disregard the messages he's sending? What should our answer be to this question?

The NASA administrator was indeed thinking outside the box when he insisted on the F, C, and B solution. However, he also added constraints that may not have been advisable. Industry recognized this and by and large responded with conventional wisdom. "You can have two out of three," they declared, and more-or-less stopped there. An out-of-the-box response, however, might be somewhat more complicated. Can you imagine what form it could take?

The answer offered here is that all systems are different and require independent as well as specialized thinking. They cannot all be "painted with the same brush." This means that some systems may indeed be amenable to accepting the three constraints of F, C, and B, but many are not. For those that *can be* built faster *or* cheaper *or* better, that goal should be accepted. In other words, new systems might well be identified *up front* into the following eight programmatic categories:

Category	Faster	Cheaper	Better
1	Yes	Yes	Yes
2	Yes	Yes	No
3	Yes	No	Yes
4	No	Yes	Yes
5	No	No	Yes
6	Yes	No	No
7	No	Yes	No
8	No	No	No

Is there an overriding consideration that serves as a criterion for the foregoing categorization? If there is, it would certainly have a lot to do with maximizing the probability of mission success. After all, a critical mission failure basically erases, sometimes in an instant, all the talk and planning about "faster, cheaper, and better."

REFERENCES

11.1   Brooks, F. P. (1995). *The Mythical Man-Month*. Reading, MA: Addison-Wesley Longman.

11.2   U.S. Department of Defense (2003). *Operation of the Defense Acquisition System*, Instruction 5000.2. Washington, DC: DoD, May 12.

11.3   Cringely, R. X. (1992). *Accidental Empires*. New York: HarperBusiness.

11.4   Nadler, D., J. Spencer, and Associates (1998). *Executive Teams*. San Francisco: Jossey-Bass.

11.5   Wellins, R., W. Byham, and J. Wilson (1991). *Empowered Teams*. San Francisco: Jossey-Bass.

11.6 Katzenbach, J., and D. Smith (1993). *The Wisdom of Teams*. New York: Harper-Business.

11.7 Scholtes, P., et. al. (1988). *The Team Handbook*. Madison, WI: Joiner Associates, Inc.

11.8 AMACOM, A Division of the American Management Associations (1980). Effective Team Building, twelve units, audio tapes. New York: AMACOM.

11.9 Eisner, H. (2002). *Essentials of Project and Systems Engineering Management*, 2nd ed. New York: Wiley.

11.10 See www.sei.cmu.edu.

Chapter 12

Perspective 8: Thinking with Pictures

Perspective 8 has to do with bringing visuals of various types into your problem definition and solution. Some years ago, in writing a book on the subject of systems engineering [12.1], I realized that I was used to constructing diagrams of various types over a long period of time. Some of these, such as the system block diagram, came to me in engineering schools. Some were learned on the job, such as software flowcharts and data flow diagrams. Whatever the source, these diagrams were part of the "tools of the trade." All of this led me to formulating a separate chapter called "Diagramming Techniques," with the introductory line "One picture is worth more than 10,000 words." This declaration is metaphorically true, so much so that it places "thinking with pictures" as one of the nine perspectives regarding how to think outside the box.

Whereas engineers, scientists, and architects have been trained, in large measure, to construct diagrams of various kinds as part of their work, perhaps other professions have not had the benefit of such training. It might therefore be somewhat foreign to some, bringing a new way of thinking into the fray. Some of the examples of this chapter attempt to deal with issues that are rather commonplace so as to demonstrate the utility of bringing diagramming methods to everyday problems. If they are applicable to such domains, one can argue, they might also be useful when dealing with large, more complex systems. This turns out to be the case as well, as a few later illustrations will demonstrate.

Managing Complex Systems: Thinking Outside the Box, By Howard Eisner
Copyright © 2005 John Wiley & Sons, Inc.

12.1   VISUAL THINKING

In 1969, Rudolf Arnheim wrote a book on visual thinking that related directly to the thinking with pictures perspective [12.2]. One of his key points, from several years of research into the matter, was that visual perception is a distinctly cognitive activity. In effect, what this means is that as we perceive the world around us through visual means, we are at the same time stimulating and utilizing our cognitive functions. In short, as Professor Arnheim asserts, perceiving and thinking are intertwined indivisibly.

As we think about a problem and its possible solution, we are using our cognitive abilities. If, during that activity, we *force* ourselves to some type of visualization that is pertinent to the problem at hand, we have the potential of enhancing our cognition. The interconnection between cognition and visual perception is helping us in terms of problem understanding and solution.

Although there is considerably more to Arnheim's treatise, including the question, "Can one think without images?", we accept his work as support for this perspective. Visualizing in various forms (e.g., images, pictures, drawings, diagrams) will enhance our ability to think outside the box.

12.2   INFORMATION FLOW

Pictures, in contrast to linear text, convey quite a lot of information. In addition to raw counts of bits or bytes of digitized megapixel images, there is the potential for conveying an overview that cannot be matched even with pages of text. For example, consider trying to describe what your house looks like to a friend. How much easier would that task be if you had just a few photographs? The pictures tell a good part of the story, and they do so at an information transfer rate that is surprisingly high.

One example of the value of a diagram is shown in Figure 12.1. In part (a) of this figure we see older version of the Department of Defense acquisition process [12.1]. This diagram shows six sequential phases:

1. Need development
2. Concept definition
3. Concept validation
4. Engineering development
5. Production
6. Operations

In contrast, part (b) of Figure 12.1 shows a current version of the acquisition management framework [12.3]. We now see five principal phases:

1. Concept refinement
2. Technology development

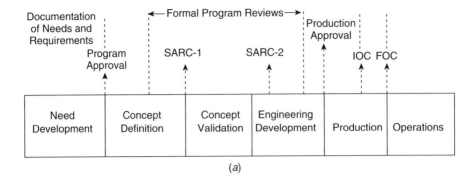

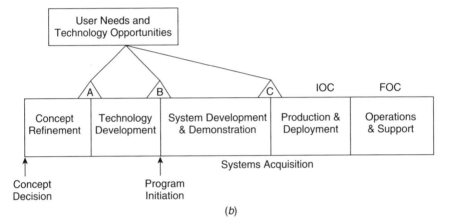

Figure 12.1 Two system acquisition processes: (*a*) defense systems acquisition process; (*b*) defense acquisition management framework (see also Figure 7.1). (*a* from [12.1], *b* from [12.3]). SARC, System Acquisition Review Council; IOC, Interim Operational Capability; FOC, Final Operational Capability.

3. System development and demonstration
4. Production and deployment
5. Operations and support

Looking directly at both parts of Figure 12.1 gives the reader an immediate sense of the two processes and of the differences between the major phases. One also notes that part (*a*) of the figure shows some reviews and milestones. By comparison, the current version [part (*b*)] places technology in a key role, since technology development and opportunities are part of the diagram. This does not mean that technology has no role in the earlier process, but it is emphasized clearly in the current process. One is able to obtain an overview rather quickly; all it takes is a relatively short review of the diagram. This is typical, even if diagrams become more complicated. However, one of the aspects of diagramming to be kept in mind is to find a level of detail that is appropriate to the reader and the application. The information flow is high, and we do not wish to interfere with the reader's interpretation and understanding with too much detail.

12.3 A DIFFERENT REPRESENTATION OF A HOUSE

Perhaps the most familiar view of a house or an apartment is what is known as the *plan view*. This is a top-down look at the various rooms, with a clear, to-scale drawing that shows the specific dimensions of the rooms. The architect knows full well how to produce this critical view, as well as the two others (elevation, perspective) that define all the necessary dimensions. An architect's product would clearly be incomplete without these representations.

However, if you are the person for whom the house is being architected, it may be that drawing a picture for the architect will help to better communicate what you want. An example of such a picture is shown in Figure 12.2. This is a type of *interface diagram* more common to the engineering of systems than to the design of a house. The picture conveys certain interconnections that you wish to have, leaving out detailed specifications that will ultimately be handled by the architect.

Starting from the front door, we see some features that are provided by thinking through pictures:

1. There is direct entry from the outside into the foyer
2. From the foyer, one can go in any one of five directions:
   a. into the living room
   b. directly into the kitchen
   c. into a bedroom
   d. into the recreation room
   e. into a hall
3. From the living room, it is possible to enter the kitchen or the dining room.
4. From the kitchen, there is a path to the living room and the dining room.
5. From the bedroom, there is a direct connection to a bath.
6. From the rec room, one can go into a den or a bath.
7. From the hall, there is access directly to two baths and three bedrooms.
8. One of the bedrooms has a connecting bath; the other bath connects only to the hall.

Why is Figure 12.2 of special interest? Because it is a specific and graphic way for the user to communicate with the designer. Indeed, its graphic form captures the text above in a concise way that is easy to read. As such, it will help the user to clarify his or her desires and may well help to make the design process shorter and

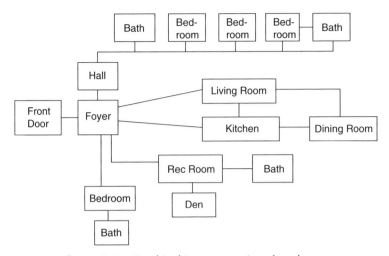

Figure 12.2 Graphical interconnections for a house.

with fewer iterations and possible false starts. Pictures can tell part of the story, often very well and with a minimum of lack of clarity or ambiguity.

12.4 GENERAL PROCESS FLOWCHART

One of the most widely used diagrams is the *process flowchart*. In its most elemental form, this type of chart simply shows sequences of activities or steps that are in series and in parallel. The difference between serial and parallel steps is illustrated in Figure 12.3. Part (*a*) of the figure shows a process with six sequential steps. One ground rule for this representation is that the preceding step must be completed before the next step can be undertaken. In part (*b*), however, steps 3 and 4 are in parallel with step 2. This means that they can be in process concurrently, but all three steps must be completed before step 5 begins. Parallel or concurrent activities are, of course, the key to shortening time lines and achieving certain efficiencies.

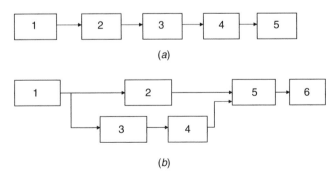

Figure 12.3 Simple serial and parallel processes: (*a*) completely serial process; (*b*) combination of serial and parallel process.

The entire field of process engineering was given a large boost with the appearance of *business process engineering (BPR)* [12.4]. As suggested in Chapter 5, BPR highlighted the process and the fact that if we were not pleased with the results we were getting, it was necessary to change the process that led to those results. This was, and is, a simple idea that was embraced by many. Along with that acceptance, numerous companies developed software that could be used to represent almost any type of process. This special capability gave us the power to change processes quickly and eventually to zero in on the best process for a given job. Today's software, of course, incorporates more complicated processes than just series and parallel steps (e.g., iterations, steps with conditions). Some of this software also fits comfortably into the category of simulation packages that are available to help in the design and optimization of complex processes such as the operation of a factory.

From the point of view of thinking outside the box, it is suggested here that the original BPR premise might be helpful in providing new solutions that are perhaps not obvious to most people. Drawing a process flowchart is a good starting point. Analyzing and changing the chart may lead to a solution that, up to that point, had been stubbornly elusive.

12.5 THE PARAMETER DEPENDENCY DIAGRAM

The *parameter dependency diagram* (PDD) is a representation of the key dependencies between variables, or parameters, in a complex system [12.5]. The first step in constructing a PDD is to develop a list of all the parameters that are considered to be of special interest in a system. Examples of such parameters for a communications system, often called *technical performance measures*, are:

1. Probability of detection
2. False alarm probability
3. Signal strength
4. Noise power
5. Detection threshold
6. Bandwidth
7. Capacity

A next step is to identify, from the list, those parameters that one might consider to be *output* variables. Then, for each of these outputs, the following question is asked:

- What does this parameter (call it P1) depend on?

As shown in Figure 12.4*a*, parameter P1 has three dependent parameters: Q1, Q2, and Q3. The same question is asked regarding Q1, Q2, and Q3. In this way, working

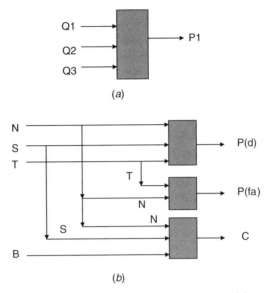

Figure 12.4 Parameter dependency diagram representations: (*a*) output parameter P1 depends on three input parameters: Q1, Q2, and Q3; (*b*) illustrative PDD for a communications system.

from right to left, we are able to construct a full PDD that shows all parameter dependencies in an explicit manner. The completed PDD sets the stage for developing a *computational road map* of the system in question.

As a further illustration of how this method works, we will construct a PDD with the specific seven parameters listed above. Referring to Figure 12.4*b*, we show the three output parameters as the detection probability, $P(d)$, the false alarm probability, $P(\text{fa})$, and the system capacity, C. These three outputs and their dependent parameters, as shown in the PDD, are listed as follows:

Output Parameters	Dependent Parameters
$P(d)$	Signal strength (S), noise power (N), threshold (T)
$P(\text{fa})$	Noise power (N), threshold (T)
C	Signal strength (S), noise power (N), bandwidth (B)

The value of the PDD is that it represents a road map of the key relationships in a system. These relationships start out as dependencies between parameters, without knowing the precise nature of each relationship. Even at the level of the PDD, we have the benefit of identifying the key variables that must be considered in greater detail as a system is designed and built. As we gain further knowledge and insight, the PDD may be augmented by converting, in a systematic manner, the dependencies into quantitative formulas. An example of the latter step is provided below for the three key output parameters cited above as well as in Figure 12.4*b*.

$$P(d) = \frac{1}{2}\left[1 - \text{erf}\left(\frac{T-S}{(2N)^{0.5}}\right)\right]$$

$$P(\text{fa}) = \frac{1}{2}\left[1 - \text{erf}\left(\frac{T}{(2N)^{0.5}}\right)\right]$$

$$C = B\log\left(1 + \frac{S^2}{N}\right)$$

The reader is urged to try the PDD method on a system with which you are familiar, or have a more direct interest. Some suggested systems include a computer, a banking system, or an automobile. To get started with the latter, consider the output parameters to be (1) time to accelerate from zero to 60 miles per hour, (2) braking distance, and (3) gas (fuel) efficiency in miles per gallon.

Embedded in the PDD is a way of thinking about problems and systems. This is one of the great values of starting with a diagram or a drawing or a picture. Through its basic structure, it draws you in a direction that may enhance your way of thinking about a problem that may be difficult to grasp.

12.6   INTERFACE DIAGRAM

We have discovered over the years that systems tend to fail at the interfaces. For that reason, we pay special attention to them, especially for large, complex systems. Even for the most complex of structures, we have found that the definition and decomposition of the functions of a system constitute an early step. Functional decomposition is also a preliminary step in the process of system architecting.

Figure 12.5 illustrates a functional decomposition, setting the stage for the additional step of finding and keeping track of interfaces between the decomposed functions. In this case, a simple matrix chart maps all the subfunctions against one another. In a first pass, the interfaces are defined as primary, secondary, or tertiary. If there is no interface of any substance, the cell is left blank.

Interfaces between subfunctions that are part of the same function are expected. Interfaces between subfunctions that are part of *different* functions are normally where problems can be found. Figure 12.5 is a starting point for dealing with system interfaces. It is then necessary to "drill down" beyond this representation in order to:

1. Identify the precise nature of the interface.
2. Try to measure the compatibility issues at the interfaces.

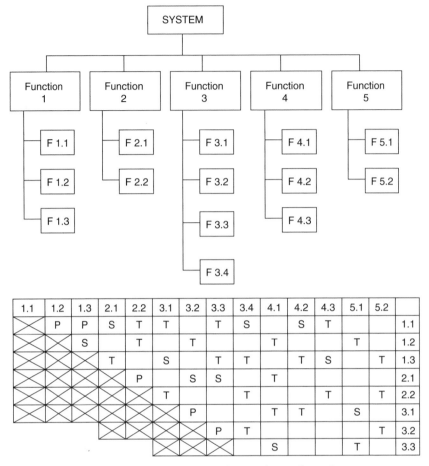

1.1	1.2	1.3	2.1	2.2	3.1	3.2	3.3	3.4	4.1	4.2	4.3	5.1	5.2	
	P	P	S	T	T		T	S		S	T			1.1
		S		T		T			T			T		1.2
			T		S		T	T		T	S		T	1.3
				P		S	S		T					2.1
					T			T			T		T	2.2
						P			T	T		S		3.1
							P	T					T	3.2
								S				T		3.3

Figure 12.5 Functional decomposition and partial interface diagram. P, primary interface; S, secondary interface; T, tertiary interface.

3. Estimate the potential risk associated with the interface.

4. Attempt to mitigate the risk when necessary.

12.7 OTHER TYPES OF DIAGRAMS

Literally dozens of diagrams can be used to assist in the process of designing and building complex systems. A short list of some that I have found to be useful is provided below [12.1]:

- Process flowchart
- Signal flow diagram

- Hierarchical decomposition diagram
- Data flow diagram
- Functional flow diagram and description
- Hierarchical input–process–output diagram
- Warnier–Orr diagram
- Michael Jackson diagram
- Action diagram
- Sequence and timing diagram
- Logic flowchart
- Nassi–Shneiderman chart
- Decision network diagram
- PERT schedule diagram
- IDEF diagram

Mastery of some of these techniques can turn out to be quite helpful in dealing with one aspect or another of designing and building large, complex systems.

12.8 ARCHITECTURAL VIEWS

The views discussed in this section refer to the architecting of complex systems, a field of immediate and current concern to both government and industry. Views of these architectures are often formulated as a picture or a diagram, under the general principle that this is the most advantageous and informative approach.

An architecting method developed by me [12.5], which is discussed in considerable detail in Chapter 13, takes the architect through a process whose main outputs are:

1. A synthesis chart
2. An analysis chart
3. A cost-effectiveness chart

All three charts are illustrated, in compact form, in Figure 12.6. The *synthesis chart* defines, for each alternative, ways of instantiating all of the subfunctions. The *analysis chart* numerically assesses the effectiveness of the alternatives, based on a set of well-defined evaluation criteria. The *cost-effectiveness chart* is a graph that plots the cost and effectiveness of each of the alternatives. These charts can also be used to draw an appropriate inference as to the nature of the architecting process itself. That is, by seeing the chart it is possible to write down the specific steps that were necessary in order to produce the chart.

		Alternative A	Alternative B	Alternative C
Function 1	Subfunction 1.1			
	Subfunction 1.2			
	Subfunction 1.3	How each subfunction is to be realized through specific design choices, for each alternative		
Function 2	Subfunction 2.1			
	Subfunction 2.2			
	Subfunction 2.3			

(a)

		Alternative A		Alternative B		Alternative C	
Evaluation Criteria	Weights	Score	Weighted Score	Score	Weighted Score	Score	Weighted Score
1. Performance							
2. RMA							
3. Security							
4. Interop.							

(b)

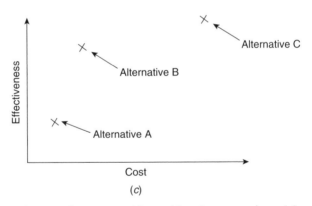

(c)

Figure 12.6 Outputs from a specific architecting procedure: (a) synthesis chart; (b) analysis chart; (c) cost-effectiveness chart. (From [12.5].)

The Department of Defense (DoD) has spent a considerable effort in dealing with the field of architecting large, complex systems. They have developed three essential views as part of their C4ISR architectural framework [12.6] (see also Chapter 8). These views are the (1) the operational view, (2) the systems view, and (3) the technical view. Although the names of these views are rather descriptive, the DoD provides many examples of how to construct such views. The architectural framework document lists the following as examples of the essential and supporting views:

All Views

Essential
Overview and summary information
Integrated dictionary

Operational View

Essential
High-level operational concept graphic
Operational node connectivity description
Operational information exchange matrix
Supporting
Command relationships chart
Activity model
Operational rules model
Operational state transition description
Logical data model

Systems View

Essential
System interface description
Supporting
Systems communications description
Systems matrix
Systems functionality description
Operational activity to system function traceability matrix
System information exchange matrix
System performance parameters matrix
System evolution description
System technology forecast
Systems rules model
Systems state transition description
Systems event/trace description
Physical data model

Technical View

Essential
Technical architecture profile

Supporting

Standards technology forecast

We take note of the fact that in virtually all cases, views of architectures are expressed as diagrams or pictures. The diagram facilitates the architecting process; the process is enhanced and expressed, in large measure, by the diagram. It is a good example of what Arnheim meant by his term *visual thinking* [12.2]. It also supports the suggestion that if you are struggling with a particularly knotty problem, take the time to draw some pictures of what you know, or what you think could possibly relate to the solution. It might take you closer toward that solution than you can imagine.

12.9 SUMMARY: A MEETING

Frank was a vice president of a systems and software engineering company. About a year ago, the company won an important contract with the federal government. There were some 50 people now working on this contract, many of them writing and building software. During the program review two weeks ago, it was determined that the effort was behind schedule by about one month. Jack, the project manager, said that the project would be back on track by the next review, in a month's time. Since the customer was invited to the next review, now scheduled for two weeks from now, Frank wanted an interim meeting to make sure that all was well. Attending the meeting, in addition to Frank and Jack, were seven lead engineers and Karen, a technical assistant to Frank.

At the meeting it was determined that the schedule problem was getting worse. Each of the lead engineers had his or her own schedule that seemed to be okay, but Jack's master schedule, a large Gantt chart, showed further slippage. The meeting became even more frustrating for Frank when Karen suggested that she might be able to resolve the issue in a favorable way but would need a couple of days to work with the individual lead engineers. Everyone agreed since that seemed to be the best course of action. Also, everyone trusted Karen, due to her strong communications skills and honesty.

Three days later, the same group convened and Frank invited Karen to present what she had found. She unrolled a rather large new master schedule and proceeded to tape it to the wall. The new schedule appeared to be complicated and long and it was clearly an output from some type of project management software package. All gathered around the new schedule as Karen began to explain.

"What I've been doing for the last couple of days", she said, "was to talk to all of the lead engineers to obtain activity and schedule data for each of our seven subsystems. Then I put all of that information into the project software and obtained the PERT chart you see in front of you. This chart shows dependencies that we had not been aware of before. It also shows explicitly the critical path, which does indeed show that without any changes we will be about seven weeks late. Then I was able to explore moving some activities off the critical path and doing several tasks in parallel instead of in series. This was checked out with the lead engineers to see if these changes were workable. It was confirmed that they were. I then put the changes into the program and produced a new master schedule, which is the one I'm now putting on the wall. This new schedule shows that we can be on schedule by making the changes that I just mentioned."

Everyone gathered around the new chart, checking to see if the new activities were acceptable. After a few minutes, all confirmed that it not only could be done but was actually a better solution for each subsystem as well as for the overall system.

"You've all done a great job solving this problem," said Frank, "and I give special kudos to Karen for understanding how to use PERT charting to get us out of some significant trouble. I thank all of you for cooperating and working the problem constructively. I guess our individual Gantt charts just didn't do the job for us. It looks like we can tell the customer that we have a plan for getting back on schedule, and we can prove it with this chart."

Many project engineers have learned, over the years, that the realities of task dependencies and a critical path can be revealed through the PERT procedure and that it is available as a feature of very inexpensive project management software. It's true that complex systems require more complicated methods for schedule production and tracking, but these methods are essential. And at the center of the capability is a diagram, one that tells the story and helps us to solve an important and real problem. This is only one of many problems whose solution can be facilitated through the use of some type of picture.

REFERENCES

12.1 Eisner, H. (1988). *Computer Aided Systems Engineering*. Englewood Cliffs, NJ: Prentice Hall.

12.2 Arnheim, R. (1969). *Visual Thinking*. Berkeley, CA: University of California Press.

12.3 U.S. Department of Defense (2003). *Operation of the Defense Acquisition System*, Instruction 5000.2. Washington, DC: DoD, May 12.

12.4 Hammer, M., and J. Champy (1998). *Reengineering the Corporation*. New York: HarperBusiness.

12.5 Eisner, H. (2002). *Essentials of Project and Systems Engineering Management*, 2nd ed. New York: Wiley.

12.6 U.S. Department of Defense (1997). *C4ISR Architectural Framework*, v. 2.0. Washington, DC: Architectures Working Group, DoD, December 18.

Chapter 13

Perspective 9:
The Systems Approach

Perspective 9, the systems approach, is appropriately the final suggestion for thinking outside the box, as it is the broadest and most inclusive. It tries to focus on the *whole* of a system, attempting to take into account interrelationships between the parts of a system, whether inadvertent or intentional. These are sometimes also called *interactions* and *interfaces*, and empirical data support the notion that rather than the devil being in the details, it's in the interactions and interfaces. Interactions that we have not fully accounted for can cause a complex system to fail, at times catastrophically. Two of our space system disasters (i.e., *Columbia* and *Challenger*), unfortunately, are testament to the fact that this can happen. The increasing complexity and size of our systems suggests that we must be even more careful in the future about interrelationships and interfaces.

We will take the point of view here that systems engineering is a formal discipline that attempts to fully utilize the systems approach [13.1]. Therefore, we will be looking at various elements of systems engineering in terms of this role. Further, some of the guiding principles of the systems approach give our systems engineers a basis for what needs to be accounted for as well as accomplished. Later in the chapter we will see how this might work.

Finally, we will use some of the ideas set forth by Senge [13.2] as a backdrop for our systems approach perspective. Recall that Senge described five crucial disciplines that need to be mastered in today's organizations:

1. Building shared vision
2. Personal mastery

Managing Complex Systems: Thinking Outside the Box, By Howard Eisner
Copyright © 2005 John Wiley & Sons, Inc.

3. Mental models
4. Team learning
5. Systems thinking

The last of these, systems thinking, is the overarching discipline that is the ultimate goal. Systems thinking allows one to proceed with the systems approach, and the latter requires the former. In addition, Senge takes one further step. He claims that to capture the five critical disciplines, it is necessary that the enterprise become a *learning organization.* Such an organization increases its chances of being successful.

13.1 SEVEN ESSENTIAL ELEMENTS

We will accept the following as essential elements of the systems approach [13.3]:

1. A systematic and repeatable process
2. Emphasizes interoperability and harmonious operation
3. A cost-effective solution to the customer's problem
4. Full consideration of alternatives
5. Uses iterations to converge and refine
6. Leads to satisfaction of all final requirements
7. Leads to a robust system

13.1.1 Systematic and Repeatable Process

Many companies have defined a systems approach that works for them. They then proceed to insist that all project teams follow that approach. Although these processes may not be the same from company to company, each company is trying to establish a systematic and repeatable process that applies to their lines of business as well as their personnel and set of customers. This notion is also supported by the various capability maturity models that define key process areas that need to be learned and adopted as permanent practices.

13.1.2 Interoperability and Harmonious System Operation

The systems approach looks at the whole system and recognizes that we tend to have problems at the interfaces, both external and internal. Hence, there is a special focus on having the parts interoperate effectively and harmoniously. Some of our very simple systems are indeed plug-compatible (e.g., commercial stereo systems), but for large, complex systems we do not have this type of assured interface. Standards in various related fields (e.g., software engineering) tend to be helpful, but they almost always lag behind the realities of new system structures.

13.1.3   Cost-Effective Solution

Seeking the most cost-effective solution is a commonsense approach to the problem. A Department of Defense (DoD) acquisition directive [13.4] accepts this perspective: "The DoD components shall seek the most cost-effective solution over the system's life cycle." At times, this solution has also been called the *best value* choice. This approach is also in consonance with how we tend to behave as consumers in the commercial marketplace. That is, we buy the product that represents the most cost-effective solution to our perceived needs. Such a solution is usually the right answer, unless there is a significant cost constraint.

13.1.4   Full Consideration of Alternatives

Full consideration of alternatives is one of the most important aspects of the systems approach. A failure to look broadly at more than one solution often leads to the wrong selection at the system level, as well as below that. This point is elaborated upon later in this chapter.

13.1.5   Iterations to Converge and Refine

One of the well-known proponents of system quality made a special point of "doing it right the first time" (DIRFT) [13.5]. For large and complex systems, doing it right recognizes that there are many unknowns and TBDs (to be determined's). Our understanding grows as we explore the system possibilities and we are better able to fill in the blanks. In the engineering of large systems we are able to combat complexity by using iterative processes that ultimately allow us to converge to the best solution.

13.1.6   Satisfaction of Final Requirements

Despite various perspectives about capability-based acquisition and variable requirements, we should not lose sight of the fact that it is necessary ultimately to satisfy the customer's needs. The pathway to achieving this may be somewhat tortuous from time to time, but it remains one of the guiding principles of building systems. The adjective "final" in front of requirements is purposeful. Even if requirements are creeping and changing, it is a set of *final* requirements that must be satisfied.

13.1.7   Robust System

A robust system is one that continues to operate under various kinds of stress, although it may be forced to do so in a degraded mode. It is a "slow-die" system that has relatively few single-point failures that are catastrophic. Such a system has been designed with special consideration given to reliability, maintainability, and availability (RMA) factors, as well as the required logistics support to assure appropriate RMA levels.

13.2   THE BREADTH OF SYSTEM CONSIDERATIONS

Since systems engineering is a confirmed means by which the systems approach is carried out, by examining the elements of systems engineering we are able to see the breadth of the required system considerations. This was done, in part, in Chapter 2, where we saw that some 30 elements constitute the basic structure of systems engineering [13.3]. These elements certainly support the notion that the systems approach is quite broad, requiring knowledge of many factors that are part of building large, complex systems. However, even these 30 elements do not tell the entire story. Following is a short list of some additional considerations, including some well-known *attributes* of the systems we are building:

1. Environmental effects
2. Sustainable development
3. User-friendliness
4. World-friendliness
5. Interoperability
6. Maintainability
7. Upgradability
8. Technology insertion and related risk
9. Obsolescence
10. Overall cost-effectiveness

If we are to be able to employ the systems approach, we need to try to be sure that something important is not being overlooked. At the rapid pace of today's world, this is not a simple task, since the problem areas tend to be quite expansive. Tomorrow's world is likely to require a response that is both broad and deep.

13.3   THE PERSISTENCE OF ALTERNATIVES

As if the breadth of the systems approach were not enough, we now have one of the key principles to consider: that of looking at *alternatives* wherever and whenever possible. In our search for the best ways to build and manage complex systems, we are immediately struck by the fact that there are many alternatives at almost every turn. This should not come as a surprise if we consider our role as consumers. In this capacity we regularly look for alternatives in search of the product that best meets our needs at a reasonable price. Whether we are purchasing a house, an automobile, a computer, or a TV set, we are aware that alternatives exist and that part of the task is to examine and explore alternatives. Further, we are suggesting that the search for alternatives should be standard operating procedure when we are building and managing large, complex systems. The same old approach that worked yesterday may not work tomorrow. If to the person with a hammer everything looks like a nail, what happens when only threaded shafts are available?

The regular search for alternatives is also part of Senge's learning organization, and vice versa. Looking beyond the standard solution, otherwise known as thinking outside the box, stretches our engineers and scientists. This process of stretching requires learning as part of it. Learning about new solutions creates alternatives. Creating alternatives means that we must constantly be learning. Alternatives and learning are mutually supportive and mutually beneficial.

One example of the search for alternatives, on a large scale, can be seen in the government's approach to building what was called the *joint strike fighter*. Both Lockheed Martin and Boeing were invited to present their designs, constituting two major alternatives for the government to consider. Eventually, Lockheed Martin was selected and we can assume that by actively encouraging alternative designs, the country is better off in terms of eventual results. The formulation of alternatives is an explicit part of my approach to architecting systems. The way in which this is carried out will be made specific very soon.

13.4   BUILDING A SYSTEM

An example of using the systems approach to construct a system is shown in Figure 13.1. Here we have a flowchart delineating the general steps necessary to take us from a set of user needs and requirements to a fully built and tested system. We normally start with user needs and requirements that are derived from those needs (box 1). We then synthesize several alternatives (box 2), using a range of technologies (box 3). These technologies help us to visualize systems of increasing

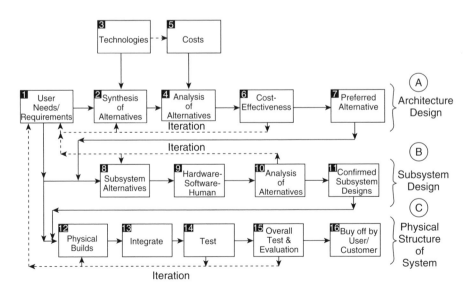

Figure 13.1   Basic steps in building a system.

performance and, usually, increasing costs and risk. The synthesized alternatives are then subjected to an analysis (box 4) that serves to evaluate the cost and effectiveness (box 6) of the alternatives. The cost estimates (box 5), in part, are determined from the technologies we have chosen to be part of the system. After some iterations, we are then in a position to select a preferred alternative (box 7). This selection completes the *architectural design* of the system (shown as "A").

From there we enter into a *subsystem design* phase, which begins by identifying the key subsystem design alternatives (box 8). At the subsystem level we are able to define roles that depend on the selection of hardware, software, and human solutions (box 9). These alternatives are subject to analysis (box 10), whose purpose is to confirm our designs at the subsystem level (box 11).

When this is complete, we have a satisfactory design at the subsystem level. These two distinct levels of design (architectural, subsystem) give us the information that we need to proceed into the actual building of the system, which begins with various hardware and software builds (box 12). These builds are integrated (box 13) and tested (box 14). When the builds have been integrated and tested to the point at which the entire system has been constructed, the system is subject to a key step of test and evaluation (box 15). If the system passes test and evaluation, there is normally a buyoff by the customer, and the system is ready for formal purchase.

It should be noted that Figure 13.1 is a somewhat idealized version of the overall process leading to a system that the customer finds to be acceptable. There are many times when one wishes to begin subsystem design before the architectural design has been completed, and also cases when parts of the system are built even before architectural design has been achieved. An example of the latter is when *rapid prototyping* is undertaken. This action is appropriate when one wishes to reduce risk by trying to solve a particularly difficult problem. In other words, even though Figure 13.1 shows three distinct phases, there can be, and usually is, a lot of work that is overlapping, for good and sufficient reasons. We note that the design process itself has two parts, one dealing with high-level choices (the architecture) and the other focusing at the subsystem level. This can be seen to be, by analogy, what is done in an A & E (architectural and engineering) firm, where the distinction between architecting and engineering is more sharply defined and accepted. Now let us look more deeply into the details of how the architecting is accomplished. This is a area of design that is recognized as important but has been somewhat controversial with respect to finding a process that will guarantee that an appropriate architecture will emerge as a result.

13.5   ARCHITECTING A SYSTEM

A formal method by which we architect systems is an integral part of the systems approach as well as systems engineering [13.3]. As depicted in Figure 13.1, architecting is viewed as a critical design process that is followed by the two important steps of subsystem design and the physical construction of the system. Although architecting has been discussed broadly in several earlier chapters, it is in the context of the systems approach that a more definitive view of architecting is presented here. This is accomplished by using a specific illustrative system and four well-defined steps:

1. Functional decomposition
2. Synthesis of alternatives
3. Analysis of alternatives
4. Cost-effectiveness graphic

Step 1: Functional Decomposition

The architecting example presented here is for a severe climate anemometry system (SCAS) [13.6]. The top-level functional decomposition for this system is as follows:

1. Atmospheric sensing
   1.1   Wind speed sensing
   1.2   Wind direction sensing
   1.3   Pressure sensing
2. Mechanical service
   2.1   Instrument housing
   2.2   Orientation/position
3. Environmental service
   3.1   Ice control
4. Power service
   4.1   Main power supply
   4.2   Power regulation/conditioning
   4.3   Backup power
5. Indoor/outdoor transmission
   5.1   Power transmission
   5.2   Signal transmission
6. Data handling
   6.1   Data collection
   6.2   Data processing/storage
   6.3   Reporting, distribution, and display

We note that a key purpose of the overall system is to measure wind speed and direction as well as barometric pressure. This type of functional decomposition identifies *what* functions are to be part of the system but does not as yet address *how* this is to be accomplished. That is reserved for the next step.

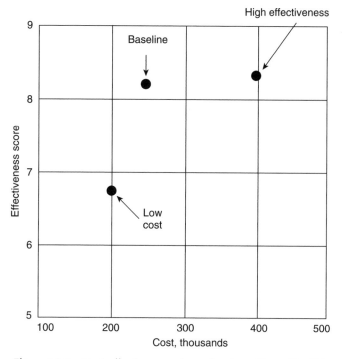

Figure 13.2   Cost-effectiveness view of system. (From [13.3].)

Step 2: Synthesis of Alternatives

Given the functional decomposition, the architect now constructs a table of the ways in which each function and subfunction is to be instantiated, for at least three alternatives (see Table 13.1). As discussed previously (see Chapter 5), the three basic alternatives are (see Figure 13.2):

1. The low-cost alternative
2. The knee-of-the-curve or baseline alternative
3. The high-performance alternative

In Table 13.1 we see the specific design concepts for all subfunctions. As an example, and with respect to the measurement of wind speed, the architect shows three alternatives. They range from a simple COTS pitot tube, to one with a transducer, to another with a radio (wireless) transducer. This process is repeated for all subfunctions, constructing the table from left to right, one row at a time. After the overall chart has been developed, each system is checked from top to bottom (down a column), to assure interoperability among the design approaches for all subfunctions. This last important step helps to assure that the individual design choices will "play well together" (in harmonious and compatible operation).

TABLE 13.1 Alternative System Architectures for Anemometry System

Functions		Subfunctions	Low Cost	Baseline	High Effectiveness
1. Atmospheric sensing	1.1	Wind speed sensing	COTS pitot tube	COTS pitot with transducer	Add radio transducer
	1.2	Wind direction sensing	Simple shaft drive	Simple shaft drive	Simple shaft drive
	1.3	Pressure sensing	COTS pitot tube	COTS pitot with transducer	Add radio transducer
2. Mechanical service	2.1	Instrument housing	Machined aluminum	Add molded composites	Less weight/compact
	2.2	Orientation/position	Wind-vaned COTS bear	Less tail boom length	High-precision bear/balancing
3. Environmental service	3.1	Ice control	Analog feedback temperature control	Add digitized control	Add process and heat pipes
4. Power service	4.1	Main power supply	Commercial 220/110 V COTS	Commercial 220/110 V	Commercial 220/110 V
	4.2	Power regulation/ conditioning	Conditioners/rods	Add ground-fault interrupter	Add lightning arrester
	4.3	Backup power	Battery instruments	Gas generator with sensor	High-reliability diesel with switch
5. Indoor/outdoor transmission	5.1	Power transmission	Stranded wire harness	Stranded wire harness	Custom slip rings
	5.2	Signal transmission	Foil-shielded wire harness	Coaxial with slip rings	Two-way radio, no wiring
	5.3	Physical linkages	Shaft/conduit, pressure tube	Add shielded transducer	Minimum shaft-physical support
6. Data handling	6.1	Data collection	Potential and indoor pneumatic cell	Magnetic position sensor	Optical position sensor
	6.2	Data processing/storage	Manual database entry	Automatic computer control	Automatic computer control
	6.3	Reporting, distribution, and display	Physical meters manual	GUI + modem access	DBMS + packet network

Source: [13.3].

TABLE 13.2   Evaluation Framework for Architecting Illustrative System

Evaluation Criteria	Weights	Low Cost		Baseline		High Effectiveness	
		Score	Weight × Score	Score	Weight × Score	Score	Weight × Score
Performance	0.3	6	1.8	8	2.4	9	2.7
Human factors	0.2	7	1.4	8	1.6	9	1.8
Maintenance	0.2	7	1.4	9	1.8	9	1.8
Risk	0.2	8	1.6	8	1.6	6	1.2
Other	0.1	6	0.6	7	0.7	9	0.9
Sums	1.0		6.8		8.1		8.4
Costs of alternatives			$200,000		$250,000		$400,000

Source: [13.3].

Step 3: Analysis of Alternatives

The next step is to evaluate and compare the effectiveness of the three alternatives quantitatively. This is demonstrated for the same SCAS in Table 13.2. A set of formal evaluation criteria and weights for these criteria are defined. In this example, the following criteria are used:

1. Performance
2. Human factors
3. Maintainability
4. Risk
5. Other

Weights are estimated and normalized to the value 1.0. These criteria, which measure the effectiveness of the systems in question, are derived from the requirements documents as well as from good systems engineering practice. The alternatives are rated on a scale from 1 to 10, and the final scores are a summation of the weights times the ratings. In this example, we see the following effectiveness scores:

low-cost alternatives	6.8
baseline alternative	8.1
high-performance alternative	8.4

Step 4: Cost-Effectiveness Graphic

Despite the fact that the effectiveness scores above favor the high-performance system, we cannot select that as our preferred system until we include cost information and consider the alternatives on a cost-effectiveness basis. As the next step in the process, then, we calculate the life-cycle costs of the three alternatives. We then treat cost as an independent variable (CAIV in DoD parlance) and plot the results, as shown in Figure 13.2. This graph demonstrates the knee-of-the-curve notion such that:

1. We obtain a large improvement in system effectiveness per dollar in moving from the low cost to the baseline system.
2. The effectiveness per dollar levels off dramatically in moving from the baseline to the high-performance system.

Therefore, the baseline system might be said to provide the *best value* to the customer as distinct from the low-cost and high-effectiveness alternatives.

The discussion above is the essence of the architecting process suggested here. In summary, it is characterized by the following features:

1. Top-down functional definition and decomposition
2. Explicit definition of alternatives
3. Separation of functions from means to instantiate functions
4. Quantitative measurement
5. A cost-effectiveness bottom line
6. Repeatable and systematic process
7. Explicit use of evaluation criteria to measure effectiveness and reflect key system requirements
8. Framework for sensitivity and trade-off analyses and studies

The "bare-bones" description above conveys the basics of this architecting process. For additional information, the reader is referred to my textbook on the subject [13.3].

13.6 ADDITIONAL VIEWS OF ARCHITECTURES

The architecting method described here focuses on the four steps described previously by example. Each step provides a specific output. The first step is the construction of a hierarchical decomposition diagram. The next two are formatted as tables or spreadsheets, and the last is a graph of the system effectiveness vs. life-cycle cost. Called *views* of the architecture, these take a central position in a quantitative understanding of the competing systems. However, having constructed these views, it is now possible as well as desirable to consider additional views that will help the decision maker select the preferred architecture. Following are 10 additional views that measure important features of the alternative systems [13.7]:

1. Requirements satisfaction
2. Risk and requirements
3. Interoperability
4. Cost by function

TABLE 13.3   View of Risk and Requirements

Function1.1	Risk of Not Meeting Requirement		
	A	B	C
Req't R1		▨	▨
Req't R2	▨	▨	▨
Req't R3		▨	▨
Req't R4	▨	▨	▨
Req't R5	▨	▨	▨

TABLE 13.4   View of Estimated Cost by Function

Function	Estimated Cost (millions of dollars)		
	A	B	C
1.1	2.0	2.5	3.2
1.2	1.8	2.3	4.1
1.3	3.2	3.6	5.0
2.1	2.4	2.8	3.7
2.2	2.6	3.3	4.8

5. Cost vs. requirements
6. Sensitivity to weight changes
7. Effectiveness vs. risk
8. Effectiveness vs. human factors
9. Effectiveness vs. RMA (reliability–maintainability–availability)
10. Effectiveness vs. performance factors

Tables 13.3 and 13.4 provide examples of two of the views listed above.

It is also recalled that the DoD has set forth three specific views that it considers have special merit—the operational view, the systems view, and the technical view—as well as subordinate views [13.8]. These three views do not conflict with the other views suggested here. Rather, they can be considered complementary.

As a matter of general principle, defining and constructing multiple views of architectures helps to increase our understanding of the advantages and disadvantages of each alternative. In the final analysis, to decide on a solution, it is necessary to use an integrated view. Between the synthesis and analysis views presented above, we have a very close representation of an integrated perspective as to how the alternatives have been constructed as well as how they compare on a quantitative basis.

13.7 ANOTHER GOVERNMENT PERSPECTIVE

The systems approach perspective has been adopted by several agencies of the federal government. As might be expected, the precise definition of what constitutes such an approach tends to vary from one department to another. The DoD, often a leader in matters of this type, has made several definitive as well as relevant statements in its directive [13.4] regarding the defense acquisition system. As one example, the following is a quote from Directive 5000.1:

> Acquisition programs shall be managed through the application of a systems engineering approach that optimizes total system performance and minimizes total ownership costs. A modular, open systems approach shall be employed, where feasible.

A second example is a paragraph whose title is *total systems approach:*

> The Program Manager (PM) shall be the single point of accountability for accomplishing program objectives for total life-cycle systems management, including sustainment. The PM shall apply human systems integration to optimize total system performance (hardware, software, and human), operational effectiveness, and suitability, survivability, safety, and affordability. PMs shall consider supportability, life-cycle costs, performance, and schedule comparable in making program decisions. Planning for Operation and Support and the estimation of total ownership costs shall begin as early as possible. Supportability, a key component of performance, shall be considered throughout the system life cycle.

This view of the systems approach tends to emphasize the PM and the broad scope of issues that such a person needs to address and be accountable for. Whatever the precise interpretation, it is clear that the DoD fully addresses and endorses both systems engineering and the systems approach.

13.8 SUMMARY: A MEETING

Sharon was a high-powered and competent manager in charge of a project under a contract to a government agency. The project involved the tracking of personnel, including considerable amounts of information about each person in the database. The system was a large one so that the personnel of an entire administration could be included.

There were 14 people on the project, excluding Sharon. The project was starting to be late, due to the fact that there was a clock time requirement that the system was not meeting. Virtually all people on the team were engaged in trying to simplify and modify the software so that the clock time requirement could be satisfied. Another month went by, and finally, Sharon had an all-hands meeting to explore what to do.

There were 15 people in the room when Sharon expressed her deepest concern about the system not meeting specification and what she discerned as being the consequence. The latter was, in her mind, very severe because they were working on a fixed-price contract and because the customer needed the system as soon as possible. A few suggestions were made as to how to reexamine and restructure the software to make it run faster. Each suggestion was treated fairly and as being an alternative that the team would consider on the spot. Sharon indicated that she wanted the people in the room to break into four smaller groups, the purpose of which was to explore all the implications of the suggested fixes. Just before these groups started working, one of the junior software engineers, Joyce, stood up to speak.

Joyce: We are all working hard trying to find a solution from among the alternatives suggested. My sense is that all the suggestions will leave us with significant increases in both cost and schedule. I've got another alternative to put on the table. How about going back to our customer and flat-out asking for relief from the clock time requirement? Perhaps he doesn't really need such a stringent clock time response.

Sharon stopped in her tracks and one could see that she was taking Joyce's idea quite seriously.

Sharon: That's a good idea. Somehow I put it aside as if it just couldn't be done. If we can get relief on the clock time, the entire problem goes away.

A week later, after a lot of case building, Sharon and two of her senior software engineers went to see the customer to ask for relief on the clock time requirement. After some discussion about changing to a clock time spec that all were sure could be met, the customer agreed that such was the course of action he would select. So the problem went away by considering an alternative that had previously been overlooked.

Making sure that all applicable alternatives have been duly considered is an integral part of the systems approach. Joyce remembered that the customer was part of the "system." Sharon was so deeply engrossed in trying to meet the clock time requirement that she had forgotten to look beyond it—specifically, at the customer. (This scenario came from a real-world experience of mine.)

REFERENCES

13.1   Sage, A., and J. Armstrong, Jr. (2000). *Introduction to Systems Engineering*. New York: Wiley.

13.2   Senge, P. (1990). *The Fifth Discipline*. New York: Doubleday Currency.

13.3   Eisner, H. (2002). *Essentials of Project and Systems Engineering Management*, 2nd ed., New York: Wiley.

13.4   U.S. Department of Defense (2003). *Operation of the Defense Acquisition System*, Instruction 5000.1. Washington, DC: DoD, May 12.

13.5   Crosby, P. (1984). *Quality Without Tears*. New York: New American Library, A Plume Book.

13.6 This original system was architected by Richard C. Anderson, a graduate student at The George Washington University. Changes in the original architecture have been made for purposes of presentation in the format of this book.

13.7 Eisner, H. (2004). *New Systems Architecture Views*, presented at the 25th National Conference of the American Society of Engineering Management, Alexandria, VA, October 20–23.

13.8 U.S. Department of Defense (1997). *C4ISR Architectural Framework*, Version 2.0. Washington, DC: Architectures Working Group, DoD, December 18.

Chapter 14

Thinking in Groups

The preceding nine chapters have presented nine perspectives regarding possibly new ways of thinking about building and managing large and complex systems. These perspectives are the main focus of this book and are meant to apply to the individual. That is, they are suggestions for how one might think in new ways about systems matters, especially if current ways are not as successful as one might hope. The nine perspectives are thought of metaphorically as being outside the box and therefore will take some adjusting to as experience with them takes place. Trying them, as the situation calls for it, is probably the simplest approach.

We have found, however, that thinking as an individual is not necessarily the same as thinking as part of a group, or the same as what a group decides to do, given sets of inputs from its participants. Many researchers have studied group and team behavior, with many different types of group rules, dynamics, and protocols. They have also seen that group behavior and results can be quite different, depending on whether or not the group (or team) is highly functional or basically dysfunctional. These are very important topics since most of us function, one way or another, as part of a group or team. And even though we may improve our ways of thinking in line with the perspectives suggested in this book, we may find that somehow things are changed when we are part of a group. These changes may or may not be for the better. In any case, we clearly want our groups and teams to be producing high-quality processes and outcomes, especially if the participants, in the main, are producing high-quality inputs. The situation is depicted in the Figure 14.1. Here we show high-quality inputs that might derive from some of the perspectives in this book. If they are inputs to a highly productive group

Managing Complex Systems: Thinking Outside the Box, By Howard Eisner
Copyright © 2005 John Wiley & Sons, Inc.

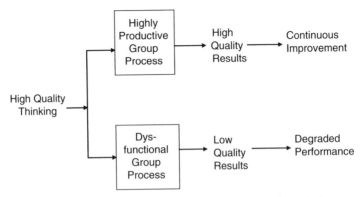

Figure 14.1 High-quality inputs can become degraded.

process, high-quality results are likely to accrue, thus supporting continuous improvement in the enterprise. On the other hand, even high-quality inputs will be degraded through a dysfunctional group process, leading to low-quality results that are likely to hurt the enterprise. The remainder of this chapter provides an overview of some of what we've learned about group and team behavior, and various ways in which we've attempted to make the group process more workable and more productive.

14.1 THE DELPHI PROCESS

One of the early suggestions for a useful group procedure was given the name the *Delphi process*, which still appears to be widely known and used. The original Delphi method was devised in the 1950s by the RAND Corporation, a company that carried out early R&D programs for the federal government. They helped to support the DoD during the McNamara "whiz kids" years, with emphasis on developing cost algorithms and estimates for a wide variety of systems. Sage and Rouse provide an excellent summary description of the Delphi process, including variations thereof [14.1].

The basic purpose of the method is to converge to a group consensus and at the same time, encourage controlled group interactions. Questionnaires are used to achieve anonymity, and several iterations are used to allow people to change their minds and thereby achieve convergence. Feedback from a facilitator is provided between rounds. A statistical statement of the views of the group is also part of the process. Of course, an attempt is made to have experts in a given field as the participants, for obvious reasons. It is desired that the totality of responses converge to one or more "best" answers. This method, with some variations, has survived over the years, probably since many believe that it is not possible to do better than a consensus of experts on matters that cannot be rigorously proven.

14.2 GROUPTHINK

The term *groupthink* has been attributed to Yale psychologist Irving Janis [14.2], who suggested that the desire for unanimity in the process of group decision making can lead to poor-quality decisions. Rather than participants speaking up and possibly voicing dissenting opinions, they are either quiet or agree when they don't really want to do so. As a result, both the decision process and the decision itself can be flawed.

Several examples tend to support this phenomenon. One deals with the January 1986 space shuttle *Challenger* and the tragedy of its massive and sudden failure. Several have conjectured, and tried to prove, that groupthink was the culprit and that the disaster could have been avoided. Another example is the Bay of Pigs invasion of Cuba, which President Kennedy conceded was a collosal mistake. Recognizing the problems in coming to such a decision, he changed the process itself so as to encourage dissenting views and the consideration of a wide range of alternatives. Indeed, it was reported [14.3] that he made five specific changes in the decision-evaluation process:

1. Insisted upon everyone's participation as skeptical generalists
2. Asked Robert Kennedy and Ted Sorenson to serve as intellectual watchdogs
3. Had task forces give up formalities
4. Called for subgroups to define and explore additional alternatives
5. He himself kept away from early meetings so as not to influence discussion and debate unduly

It may be assumed that these changes were all positive and may have had such an effect during the October 1962 "missile crisis" with the Soviet Union regarding the presence of missiles in Cuba. That difficult period, it may be recalled, worked out well, but according to many sources, we were "at the brink." Decision making by a group can thus be seen to be a critical set of activities, possibly influencing such matters as international war and peace.

A related group dynamic was articulated by J. Harvey in his well-known *Abilene Paradox* [14.4]. In his description, Harvey tells the story of four people who live in Coleman, Texas, and decide to go to Abilene for dinner at a cafeteria they know about. This is a 53-mile drive, in over 100°F heat, and in an non-air-conditioned 1958 Buick. Nonetheless, they make the trip, and when they come home they all express the view that they did not really want to go to Abilene. This paradoxical behavior is an example of what Harvey claims is the "single most pressing issue of modern organizations": that they are not really able to manage *agreement* in ways they need to. Many other examples and parables are presented to support that position in a book that is highly recommended.

Groupthink, in the above contexts, is a curious group dynamic that is important to recognize and deal with. If you have worked hard, for example, to master some of the thinking perspectives that have been suggested in this book, you might well feel the pressures of a type of groupthink at meetings that you attend. Even if your ideas are superior, if they are perceived as "far out," you might be reluctant to verbalize them. This could be a serious mistake, both for you as well as in terms of the ultimate decision that is made. If you believe that you have a new point of view that has merit or a new direction that needs to be explored, it is suggested here that you need to make it known to all in the group, with all the relevant arguments that you can think of. We are certainly in need of both the courage to do so and the potentially positive consequences that might well accrue from such action.

14.3   de BONO AND HIS THINKING HATS

Edward de Bono is one of our most original thinkers, having defined improvements in both individual and group processes. He is perhaps best known for his exposition of *lateral thinking*, which we examine in Chapter 15. For our purposes here we focus on a de Bono group process called *six thinking hats* (STH) [14.5]. Each hat is assigned a color, and each represents a thinking direction, or mode of behavior:

1. White: facts and figures
2. Red: emotions and feelings
3. Black: cautious and careful
4. Yellow: speculative and positive
5. Green: creative thinking
6. Blue: control of thinking

In a group session, people "put on" one of the hats and behave in the manner exemplified by that hat. This method has been demonstrated to be very effective in helping to solve problems and represents a workable (*preferred*, to de Bono) alternative to a more conventional "argument" approach. By focusing on only one thing at a time, in a controlled manner, each type of thinking is clarified to all that participate. This improves communication and understanding and leads to solutions that tend to be accepted by all. As stated by de Bono, his STH approach is Confucian rather than analytic.

de Bono suggests two ways in which the hats are to be used. They can be used individually to request a mode of thinking from any one person, or they can be used sequentially to examine a topic and explore its various dimensions. Discipline is required and the recommended time for a response is about 1 minute for each person. It is also suggested that the blue hat be utilized to begin and end a session. For example, in the latter case, the blue hat would summarize the following points:

1. What has been achieved
2. What specific outcomes have been produced
3. What conclusions, if any, have been reached thus far
4. What solutions have been found
5. What steps are suggested next

I have tried the de Bono STH method and found it to be highly instructive, interesting, and productive. The reader desiring additional information is referred to de Bono's book [14.5].

14.4 ADVOCACY vs. INQUIRY

A common picture of meetings in both industry and government is one in which advocates for a particular course of action argue their case. I have experienced this mode of behavior numerous times during my 30 years in industry. However, according to an investigation reported in the *Harvard Business Review*, the advocacy process is "possibly the least productive way to get things done" [14.3]. The article's authors suggest a movement from *advocacy* to *inquiry* that involves three important steps:

1. Encouraging constructive conflict
2. Ensuring that all points of view are heard
3. Knowing when and how to close deliberations

The first step brings all the key issues to the forefront so that they can be explored in great detail. Tough debates on these issues are encouraged, but they cannot be allowed to cross the line into personal attacks. The second step requires that the leader make sure that all participants are fully heard. This can mean inviting reticent players to voice an opinion and finding the right way for each person present. No participant can be allowed to dominate, and all need to participate. Finally, there is a proper time to call a halt to the proceedings, even if it means that there will be a continuation in the afternoon or next day. Mastering these three critical features of an inquiry approach requires considerable skill and sensitivity to the dynamics of the group.

Decision making via a process of inquiry leads to collaborative problem solving in which each issue is examined, tested, and evaluated in objective terms. Balanced arguments are accepted and there is an honest search for alternative courses of action. Instead of winners and losers at the conclusion, a sense of collective ownership of the results is engendered. When it works properly, there is also a sense of fairness and satisfaction for just about all of the participants. The authors report considerable success with this approach and it would appear to have considerable merit, especially for those of us who have been part of an advocacy process that has become extreme.

14.5 SAST

Strategic assumption surfacing and testing (SAST) is a group method that attempts to make assumptions explicit and then tests them for appropriateness and validity. The method is often traced back to the philosophy of C. West Churchman, a pioneer in the field of operations research. Direct attribution goes to the work of Mason and Mitroff [14.6], and four main principles are part of the procedure [14.7]:

1. *Participative.* All stakeholders should be part of the process.
2. *Adversarial.* As with our earlier discussion of *inquiry*, it is desired that adversaries raise their arguments (surfacing) so that they can be dealt with.
3. *Integrative.* Through a group process, the correct actions will be synthesized into a cohesive plan.
4. *Managerial support.* Ultimately, the plan will be supported not only by the team but by management.

The SAST approach has four phases:

1. Formation of a group
2. Surfacing of all relevant assumptions
3. A dialectical debate
4. Synthesis of options, conclusions, and courses of action

SAST supports and embodies systems thinking and has demonstrated that groups can indeed go through the four steps in a highly productive manner. A skilled facilitator is a necessity, especially to assure that the debate, and the conflict that may be engendered, does not get out of control. The crucial synthesis step shows that a cogent "summing up" can and does work if we understand that this is a key objective. Finally, by encouraging broad stakeholder participation it is more likely that a suitable and acceptable solution to a knotty problem will ultimately be found.

14.6 TEAM SYNTEGRITY

Stafford Beer, a cybernetician well known especially in the UK, developed a group process that he called *team syntegrity* [14.8]. He was looking for a process that emphasized divergent views as well as democratic principles. Curiously, he appeared to formulate his seminal idea when looking at what Buckminster Fuller had done with architectural design. What struck him in particular was the structural integrity of the icosahedron, and he thought that this design could be brought to bear in the design of the group or team process that he was looking for. That turned out to be the case, as his design led to a process with 30 people, broken down into 12 discussion groups. Each participant also belonged to two groups (right and left groups), and thus there was a total of 60 team participants.

This type of structure, for Beer, could represent a democracy with no global hierarchy. As a consequence of its structure, it also had, by analogy to the icosahedron, a type of "tensile strength" that Beer thought would provide cohesion to the overall process. The traditional form of team integrity had 30 people working over a five-day period on 12 key issues. This form was modified by others but remains the basis of Beer's original approach.

The steps in the sessions described above were five in number:

1. Definition of an opening question on which to focus
2. Formulation of a detailed agenda
3. Allocation of the 30 participants to 12 topics
4. Outcome resolution, dealing with a series of meetings at which the discussion take place
5. A closing session at which the teams present results in the form of final statements of importance

Many such sessions have been carried out over the years, yielding "very good results" [14.7]. Some of the positive statements made about the overall process have included the words: *democratic, genuine consensus, fair, multiple viewpoints, shared understanding, facilitated learning*, and *insightful agreements*. Detractors have been concerned about the rigid structure, derived as it is from the Fuller model. It certainly appears to be a unique approach from one of our leading thinkers.

14.7 FACILITATION

Getting the best results from a group can often depend, pure and simple, on the group leader or facilitator. Managers of a group may turn out to be very poor facilitators in a group situation. In such cases, groups tend not to be very good at solving problems. Conversely, managers may also be good facilitators and thereby can realize the potential synergy of positive group dynamics and behavior.

Some suggestions for how to enhance group results by focusing on the art of facilitation are provided by several authors [14.9], who also looked at the "Zen" of group behavior [14.10]. In the former, the following aspects of facilitation are examined:

1. Making sure that you can facilitate yourself
2. Facilitating a group through trust, adaptation, humor, and managing conflict
3. Using toolkits (e.g., workshops, meetings)
4. Training programs
5. Achieving synergy

For the latter, the authors, with open hearts and acknowledgments, explore in some detail a variety of topics, including:

1. The Zen philosophy and practice, and how it applies to groups
2. How groups tend to operate
3. How group behavior can change

4. Models for meetings
5. How to achieve synergy
6. An extensive articulation of tools that can be used in many situations and for many purposes

Both approaches to finding better ways for groups to behave have considerable merit and the use of tools (item 6 above) is an excellent source of very interesting information that can be used immediately.

14.8   SELF-DIRECTED WORK TEAMS

The notion of self-directed work teams sounds like a solution and a problem at the same time. The solution part is that it implies a highly functional and productive team, and that is true. The problem part is that it seems to have no boss, since it is self-directed. That, of course, is not true, although certain aspects are true. This type of team is set in motion to create a well-defined product or service for an enterprise, with a minimum of supervision, and to set an example that will engender full employee involvement.

The specific example of self-directed work teams cited here is from a book with four authors [14.11] who believe that such teams are here to stay and that they are making an important contribution to how we do business. They define a self-directed work team as "a highly trained group of employees, from 6 to 18 on average, fully responsible for turning out a well-defined segment of finished work."

The work can be a product or a service, and if the latter, the manner in which it is to be delivered is fully defined. Some differences between a more conventional group and a self-directed group are, for the latter:

1. There are only a couple of job classification categories.
2. Team decisions control the team's daily activities, without the need of a supervisor.
3. Rewards are clearly tied to the performance success of the team.

It is claimed in [14.11] that as of 1990, many companies were active in utilizing self-directed teams (a dozen are listed explicitly) and that most are reporting success stories. Some of the payoffs reported include:

1. Increases in productivity
2. More flexible and streamlined operations
3. Higher-quality products and services
4. Increased commitment to company goals, on a broader base
5. Greater customer satisfaction

With these types of results, this approach is not to be taken lightly.

It takes some time, however, for a self-directed team to become fully functional, and five stages are expected in moving from startup to maturity:

1. Startup (initial definition of team and operation)
2. State of confusion (do we know what we're doing?)
3. Leader-centered team (member of team steps forward)
4. Tightly formed team (narrow loyalties are dominant)
5. Self-directed (a mature team that is fully functional)

From the above we can see that it takes time and effort to transition through these stages successfully, and that one cannot expect instant full-up operation. To prevail, top management needs to be committed to the team-building approach and have patience as well as basic trust in both the concept and the specific people to see it through.

14.9 SYNECTICS

Apparently, the word *synectics* is derived from Greek and refers to the bringing together of different things that appear to be unrelated. It is therefore a kind of synthesis of things so as to create something new or something that had not been well recognized before. Synectics therefore deals with creativity, especially in a form that applies to groups rather than individuals, the main focus of this chapter.

The discipline of synectics appears to be largely traceable to William J. J. Gordon, who writes about the early days of this field and his related research and applications when he was with the consulting firm A. D. Little [14.12]. Gordon, along with a relatively small group of people used synectics in their work for clients as they were trying to find new products and services. Synectics evolved into a definitive process that yielded good results in this endeavor and required the exercise of high levels of creativity. A colleague of Gordon, George Prince, also wrote about those early days when the two men worked together, roughly from 1958 to 1965 [14.13]. By 1970, both men had defined in books the essential structure of synectics as well as their experiences in term of its enhancement and application over the years.

As noted above, synectics embodies a creative process whereby a group of people solve knotty problems in rather inventive and perhaps unexpected ways. It has many advantages as a group rather than an individual process and uses methods that tend to mitigate some of the problems that one finds in a group, such as general negativity, confusion, the NIH (not invented here) factor, resistance to change, internal competition, and rigidity of thinking and position. Participants are trained to recognize these types of influences and to keep away from them actively. Instead, they employ methods such as the use of metaphors and analogies to explore topics related to the problem they are addressing, giving each other lots of room to express "far-out" thoughts without being put on the defensive. Experts in needed disciplines are used so that all important information is available in the group's deliberations. The leader of a group knows when and how to pay attention to all of the interactions and become an expert at listening and asking questions without manipulating the group. Every meeting is well focused on solving the client's problem, although to the outsider, the process may appear to be roundabout. This is partially due to the belief that many solutions come from the preconscious part of ourselves, and that this part can be voiced only when an open and trusting environment is part of the process. In this sense, synectics has been associated with creative and imaginative speculation, connected to an underlying discipline that it is the job of the leader to enforce. It is also true that the leader can well be changed, from problem to problem, so that all participants in the group have the chance to function as leader.

It appears that synectics has been quite successful at solving difficult and possibly amorphous problems in a highly creative and effective manner over the years. It is often able to teach people how to think about problem solving in new ways, using procedures that they have not used before, such as creative speculation in a group setting. Apparently, the field is alive and well today since it is being practiced as a discipline by several firms that help businesses create new products and improve old ones.

14.10   NO MORE TEAMS: COLLABORATION

M. Schrage has written a most informative as well as interesting popular book on creative collaboration: what it is, why it's useful, and how it can be achieved [14.14]. His thesis, in part, is that use of the word *team* when referring to groups does not address the most important part of the story: collaboration.

Collaborations have their greatest relevance to complex problems and are therefore of great interest in terms of the central application area of this book: building and managing complex systems. We all know intuitively that success in this endeavor will be made more likely if key collaborators are present to try to deal especially with matters of complexity. Concurrent engineering addresses the issue at the top level; collaboration sharpens the focus and works on the core of the issue.

Schrage explores the matter of collaboration from many perspectives, including:

1. Conceptual and technical
2. The use of shared spaces
3. Tools that can help
4. Desires and constraints
5. The ecology of meetings
6. A wide variety of examples

Under item 6, Shrage examines in some detail collaborations between Watson and Crick, Jobs and Wozniak, Mitch Kapor (of Lotus fame) and Jonathan Sachs,

Lennon and McCartney, Braque and Picasso, and others. There are innumerable lessons to be learned about collaboration simply by hearing about what various duos have to say about their own experiences in this role.

The author also identifies very specific design themes for the reader who wants to know how to actually set up a collaborative environment. Selected areas of interest in this regard are:

1. Establishing goals
2. Engendering a feeling of trust and respect
3. Appropriate communications
4. Responsibilities and boundaries
5. The use of outsider inputs

Finally, the author provides an extremely helpful user's guide for moving ahead with a computer-supported collaborative environment. This tells us clearly how to approach the nuts and bolts of supporting the practical activities of collaborative teams.

14.11 CONCLUSION

From the overview of group and team dynamics in this chapter we can see that groups can be enhancers or they can be detractors. They can provide a clear value-added capability to the ways in which we build and manage complex systems. On the other hand, groups can degrade decision making even when individual participants come to the table with good ideas and excellent thinking. It is no surprise, therefore, that so much attention is being paid now to how to convert groups into highly functional teams and to convert gab sessions into productive meetings.

In terms of what to look out for, and try to overcome, Martino provides us with the following areas of special concern [14.15]:

1. Misinformation accepted as true
2. Social pressure that causes negative effects
3. A vocal majority that hogs the agenda
4. Drive to agreement instead of the right answer
5. A single person who subverts the team process
6. Special vested interests that are contrary to the group agenda
7. A premature leap to an unwarranted conclusion (e.g., a technology of special interest)

With respect to suggestions for building a project team and watching out for team busters, the reader may wish to consult my lists and related discussions of these important areas [14.16]. Finally, Sage and Rouse [14.1] give us a brief overview of brainstorming as well as brainwriting and groupware, the latter representing software that supports group analysis and decision-making processes.

REFERENCES

14.1 Sage, A., and W. Rouse (1999). *Handbook of Systems Engineering and Management.* New York: Wiley.

14.2 Janis, I. (1982). *Groupthink*, 2nd ed. Boston: Houghton Mifflin.

14.3 Garvin, D., and M. Roberto (2004). "What You Don't Know About Making Decisions" in *Harvard Business Review on Teams That Succeed.* Boston: Harvard Business Review Paperback.

14.4 Harvey, J. (1988). *The Abilene Paradox and Other Meditations on Management.* New York: Lexington Books.

14.5 de Bono, E. (1985). *Six Thinking Hats.* Boston: Little, Brown.

14.6 Mason, R., and I. Mitroff (1981). *Challenging Strategic Planning Assumptions.* Chichester, West Sussex, England: Wiley.

14.7 Jackson, M. (2003). *Systems Thinking: Creative Holism for Managers.* Chichester, West Sussex, England: Wiley.

14.8 Beer, S. (1994). *Beyond Dispute: The Invention of Team Syntegrity.* Chichester, West Sussex, England: Wiley.

14.9 Hunter, D., A. Bailey, and B. Taylor (1995). *The Art of Facilitation.* Cambridge, MA: Da Capo Press, Perseus Books Group.

14.10 Hunter, D., A. Bailey, and B. Taylor (1995). *The Zen of Groups.* Cambridge, MA: Fisher Books, Perseus Books Group.

14.11 Orsburn, J., L. Moran, E. Musselwhite, and J. Zenger (1990). *Self-Directed Work Teams: The New America Challenge.* Homewood, IL: Business One Irwin.

14.12 Gordon, W. (1961). *Synectics: The Development of Creative Capacity.* New York: Harper & Row.

14.13 Prince, G. (1970). *The Practice of Creativity.* New York: Collier Books.

14.14 Schrage, M. (1989). *No More Teams!* New York: Doubleday Currency.

14.15 Martino, J. (1972). *Technological Forecasting for Decision Making.* New York: American Elsevier.

14.16 Eisner, H. (2002). *Essentials of Project and Systems Engineering Management*, 2nd ed. New York : Wiley.

Chapter **15**

Widening the Circle

This book has as its centerpiece nine perspectives for thinking in new ways about large, complex systems and related problem areas. These nine perspectives are presented and discussed in some detail in Chapters 5 through 13. The perspectives have been defined by me based on firsthand use and observations of the results of that use. That is, this list is my "top nine" group of suggestions for how to tackle especially difficult management and systems problems and thereby achieve consistently positive results. Indeed, the motivation for writing the book was to attempt to bring some new approaches to the table; approaches with a history of success that could be used by others to their advantage.

However, it is also recognized that other approaches are suggested by various practitioners and teachers. For the sake of completeness, in this chapter we address some of these other approaches so that readers can explore them if they wish to do so. Of necessity, the treatment here is quite limited, serving primarily as a pointer to another source or reference. The pointer is mainly informational and is not meant to be a definitive endorsement, unless so stated. This also does not imply a negative view from me. Rather, it suggests that insufficient time has been spent on the technique, approach, or method. We now present several additional and brief citations that might prove to be useful to the reader.

15.1 de BONO AND LATERAL THINKING

Edward de Bono has been both highly creative and persistent in his formulation and explication of the notions of lateral thinking [15.1, 15.2]. This field is clearly

Managing Complex Systems: Thinking Outside the Box,   By Howard Eisner
Copyright © 2005 John Wiley & Sons, Inc.

de Bono's brainchild and his own thinking about it has ranged far and wide. For example, he has treated this type of thinking in its basic form as well as its application to management issues and problems.

de Bono is one of a kind, having been trained in and achieved competence in such fields as medicine, physiology, and psychology. He has been a prolific writer and purveyor of ideas about thinking, technology, business opportunities, and management. First and foremost, de Bono distinguishes between vertical and lateral thinking. Vertical thinking, which clearly has its place, is logical thinking that tends to point to a particular solution to a problem and then drill down in greater depth to explore all the ramifications of that solution. Lateral thinking moves sideways to generate a range of alternative solutions. The result of the lateral thinking is a broad set of alternatives, each of which can then be subjected to deeper vertical thinking. With vertical thinking only, it is often the case that only one solution is seriously considered, leading to limited problem solving and decision making.

de Bono explores and explains his notions of lateral thinking (LT) in great detail in many books. For example, he tells us that LT helps us to be open and receptive to change, rejecting many of the restrictions of vertical thinking. In addition, he provides his readers with tools and techniques that assist in becoming proficient in LT. Two processes and three methods are especially helpful [15.2]:

1. The *processes* of (a) escape and (b) provocation
2. The *methods* that deal with (a) attitude, (b) techniques and skill, and (c) the new operational word he calls "PO"

It is relatively easy to see how de Bono's LT might be applied to building and managing large complex systems by helping teams with the following:

1. Brainstorming sessions in order to formulate new ideas
2. Using new ideas to be more innovative as an enterprise
3. Finding new alternatives and solutions that otherwise might have been overlooked
4. Forcing change in the restrictive thinking patterns of some executives and managers

de Bono claims that LT is difficult to learn. Despite such a declaration, this author believes that exploring de Bono's ideas is well worth the journey.

15.2   TRIZ

TRIZ is a powerful problem-solving methodology. It comes from a Russian abbreviation that can be expressed as a *theory of inventive problem solving* [15.3]. It was originally formulated by Genrich Altshuler, who died in 1998 but left TRIZ as a legacy to be adopted and advanced by others. Its popularity has grown in the 1990s and into the twenty-first century as more people have discovered it. The types of problems that TRIZ can help one to address are:

1. Improvements in quality and quantity (contradiction problems)
2. Investigations of shortcomings (diagnostics)
3. Reductions in cost (trimming)
4. New uses of existing procedures, processes and systems (analogy)
5. New combinations of known elements (synthesis)
6. Response to a new need through reformulation (genesis)

The key reference cited here for TRIZ can be considered a handbook for a general methodology of solving technical problems. The author indicates that it is not a trivial matter to be able to learn and apply TRIZ effectively. It needs to be studied and mastered and tried multiple times. This is not unlike any definitive method that rises to the heights as does TRIZ.

TRIZ is formally defined as a "human-oriented knowledge-based systematic methodology of inventive problem solving" [15.3]. The main building blocks of TRIZ are contradiction, evolution, resources, and ideal solution.

Patent and other technical information serve as source data for TRIZ and studies thereof have led to a collection of heuristics that support TRIZ. However, if a problem is especially difficult, the methodology allows one to go to other formal structures that are part of TRIZ. One such feature is an internal *algorithm for the solution of inventive problems*, known as ARIZ. ARIZ focuses on solving non-typical problems through TRIZ heuristics. The overall structure of TRIZ may be said to contain the following main topics:

1. Preliminary analyses
2. Contradiction matrices
3. Separations principles
4. Substance-field analyses
5. Standards
6. ARIZ
7. Agents method

Although this brief summary identifies some of the key elements of TRIZ, considerable elaboration is required to properly represent the essence and value of TRIZ. For this purpose, the reader is referred to the reference cited as well as the Internet, including the URL of www.jps.net/triz.

15.3 THINKING LIKE LEONARDO

Michael Gelb's book on thinking processes attributable to Leonardo da Vinci explains "seven steps to genius every day" [15.4]. These are certainly very useful steps and worth noting in the context of this book. The seven steps are presented in the Italian language loosely translated by me as:

1. Curiosity, to achieve continuous learning
2. Committing to testing knowledge through experience
3. Use and refinement of the senses
4. Accepting paradox and ambiguity
5. Combining science and art through whole-brain thinking
6. Cultivating fitness and grace
7. Embracing the world's interconnectedness through systems thinking

Gelb provides numerous interesting and useful examples of each of the above, developing a convincing case for their overall utility. One is not only left with a feeling that these steps have been tried in a constructive manner, but also left in awe of the scope and depth of Leonardo in the context of the time period during which he made his contributions (the sixteenth century). In his words, Gelb takes a "flight through history's loftiest mind." That lofty statement is not unreasonable with respect to this unique person.

15.4 THE ART OF PROBLEM SOLVING

A focus on problem solving is almost always connected to suggesting new ways of thinking. We are interested in the latter since we wish to be better at solving problems. One of the more entertaining perspectives on problem solving is provided by Russ Ackoff [15.5], who accompanies his major subject with his "fables" that illustrate his various points. His overall thesis is that problem solving can be made to be more creative by putting art into it. In general, the decision-maker explores the following in his or her approach to problem solving:

1. Desired outcomes (objectives)
2. Possible courses of action (among variables that are controllable)
3. The environment (reflecting uncontrollable variables)
4. Relationships among all of the above
5. Constraints that may apply to one or more of the above

Ackoff takes us through these problem-solving considerations with a liberal use of parables, morals, and stories. The "art" suggests development of one's aesthetic sensibilities, in addition to the more analytic approach involving the above five elements. He also shows us how a problem-solving system may be illustrated in diagrammatic form. This is quite like a process flowchart discussed in Chapter 12.

Ackoff also places problem solving within the context of managing an enterprise by identifying what he considers to be the key elements of good management:

1. Concern
2. Competence
3. Communications
4. Courage
5. Creativity

The last of his "5 C's" elevates problem solving to where it needs to be in order to be effective. He also comments on the dangers of oversimplification (remember K.I.S.S.), as it may lead to missing or neglecting what might, upon deeper reflection, turn out the be a useful solution.

15.5   EINSTEIN

There are at least two ways of trying to understand some of the thinking patterns of Albert Einstein, perhaps the preeminent scientist of the twentieth century. One way is to read his direct writings. A second is to examine what others have said about him. Both are briefly explored here.

In Einstein's testimonial in regard to a survey on the psychology of invention in the field of mathematics by Jacques Hadamard, he comments on the latter's interest in how mathematicians think and invent [15.6]. He rejects the "words or the language" in favor of clear images that can be formed about the issue in question. Further, these images are taken in various combinations and are supported by what one might call *logical thinking*. With respect to thinking outside the norm, another quote is his belief that few people can express views that "differ from the prejudices of their social environment." Indeed, he asserts, it is difficult even to form such opinions.

Einstein showed great respect for the scientists that came before him, such as Galileo and Newton. He also showed special interest in using a *minimum number* of primary concepts and relationships to develop a new construct or theory. Further, although the world would certainly call him a theoretical physicist, he advised that all scientists should always stay in touch with their experiences. The realities of experience serve to ground the investigator, although many theories could not be immediately supported by direct experience and observation at the time the theories are developed.

One clear supporter of how Einstein thought wrote a book [15.7] that summarized his perceptions of some of Einstein's main patterns of thinking. His claim was simply that one principle stood out: being completely comfortable with *breaking the rules.* He then proceeded to demonstrate how Einstein did exactly that in many situations. Looking at history, of course, it seems that our original thinkers were all able to question the rules of their day and then to develop new rules (theories) that were ahead of their time, breaking new ground for others to build upon.

15.6 BREAKTHROUGH THINKING

The *Harvard Business Review* assembled a group of articles representing how to do *breakthrough thinking* [15.8]. Eight articles were selected for this book, looking at the topic from several perspectives. The one selected for a brief overview here is by guru Peter Drucker on matters dealing with innovation [15.9].

One of the motivations for attempting to move outside the box in one's thinking has to do with helping an enterprise be more innovative than its competitors. We have seen many instances in which a company lost its competitive edge (e.g., Wang Labs) or virtually gave it away (Xerox Management in relation to Xerox PARC) either by failing to be continuously innovative or by not paying sufficient attention to its innovators. So thinking and innovation go together, as one might expect. It also seems as if a random cross section of bosses might well be likely to be complaining about how their reports are not sufficiently innovative in how they solve problems or how they simply do their jobs.

Innovation is a complex subject, and much has been written about it. But we can scarcely do any better than to listen to what Peter Drucker has to say, in very brief form. With respect to innovation, which we can take as at least supported by out-of-the-box thinking, Drucker suggests that there are four sources of innovation opportunity within an enterprise [15.9]:

1. Unexpected occurrences
2. Incongruities
3. Process needs
4. Industry and market changes

Further, three addition sources of opportunity are outside an enterprise:

1. Demographic changes
2. Changes in perception
3. New knowledge

Drucker elaborates in some detail regarding each of these opportunity sources. However, rather than "grandiose ideas," he claims that "what innovation requires is hard, focused, purposeful work." He continues to clarify: "If diligence, persistence and commitment are lacking, talent, ingenuity and knowledge are of no avail." This perspective is well worth remembering. The wonderful breakthrough idea is necessary, but it will certainly not carry the day to product or service realization.

Another excellent example of breakthrough thinking is the result of research carried out by Nadler and Hibino [15.10]. Much more than a single article on the subject, these authors define and demonstrate a series of seven principles that have been shown to enhance breakthrough thinking:

1. The uniqueness principle
2. The purposes principle

3. The solution-after-next principle
4. The systems principle
5. The limited information collection principle
6. The people design principle
7. The betterment timeline principle

The authors are then able to show how holistic problem solving can evolve from an effective coordination of these seven principles. This is certainly an important set of perspectives regarding improvements in one's approach to new ways of thinking.

15.7 CREATIVITY

As briefly discussed above, notions of innovation are also closely related to the subject of creativity. Enterprises are interested in building and managing complex systems and in having their people take on all the attendant tasks as creatively as possible. An example of a seminal piece of work in this arena was carried out by a professor of psychology at the University of Chicago (we will call him Professor C, for reasons that can be surmised from the related reference) [15.11]. Professor C interviewed more than 90 people that exhibited special creativity characteristics. He analyzed their responses and pulled together an unusual but very striking set of conclusions. He was, of course, looking for their views on the makeup of creativity and how it can be achieved and enhanced in all our lives. He was also exploring ways that people think and work in a very broad systems context. From that perspective, he asserted that creativity could be examined only by looking at the interrelationships between three main parts of a system:

1. The domain
2. The field
3. The individual person

In the context above, Professor C defines *creativity* as "any act, idea or product that changes an existing domain, or that transforms an existing domain into a new one." In looking for creativity in a person, we are also searching for (1) curiosity and openness to problems and situations as well as (2) an "almost obsessive perseverance." These two traits can be in conflict, but when they are working together, we tend to observe at least some significant level of success.

Professor C, having spent considerable attention on a variety of matters with respect to creativity, focuses a complete chapter on enhancing personal creativity. His starting point is stated as: "So the first step toward a more creative life is the cultivation of curiosity and interest, that is, the allocation of attention to things for their own sake." If this condition is satisfied, along with extreme perseverance, the results are likely to be favorable. Professor C gives the reader quite a few specifics about what to think about and do every day to bring creativity closer. This well-written and well-conceived book [15.11] is an important part of the literature on this important topic.

15.8   COGITO ERGO SUM

A book about ways of thinking could not be complete without at least some reference to René Descartes, from whom we have gotten the now famous declaration: *Cogito ergo sum*—"I think, therefore I am." In later words of Descartes [15.12], he took these words as "the first principle of the philosophy I was seeking." This led him to conclude that what could be perceived clearly and distinctly could be taken as true, and not to accept anything as true unless this condition was satisfied. At the same time, he was able to follow doubtful opinions when sure opinions were not available, arguing that a person lost in a forest was better off following a distinct direction rather than the possibly unhappy alternative of going around in circles.

An interesting precept enunciated by Descartes was to divide a set of "difficulties" into as many parts as possible so as to facilitate the finding of a solution. This "reductionist" approach is in consonance with our overall approach to systems analysis and engineering. It may fail at times, but by and large this approach succeeds. We can decompose a system into functions and subfunctions and then look for ways to instantiate each of the subfunctions. In this process we assume that the decomposed pieces can be put together in sensible ways, and this turns out to be mostly true for both hardware and software. It would seem that Descartes, through the precept cited, was basically doing more or less the same thing.

Of course, these few notions are a very small sampling of the many thoughts set forth by Descartes, especially in his *Discourse on Method* [15.13]. The reader is urged to address this seminal text for a full representation of the way in which this very special person thought about life, nature, and thinking.

15.9   THINKING ABOUT THE FUTURE

Jennifer James has been advising us not only on how to think but also on what to think about [15.14]. The latter involves a most serious focus on the future and what it might hold for us. An activist approach to building a more holistic future means that eight skill areas will need much greater development and maturity:

1. Perspective
2. Pattern recognition
3. Cultural knowledge
4. Flexibility
5. Vision
6. Energy

7. Intelligence

8. Global values

On the matter of new ways to think, James cites four styles that people use to process information:

1. Analytical

2. Conceptual

3. Structural

4. Social

Many of us are able to use more than one of these styles rather effectively. Utilizing teams for problem solving is more likely to bring all of these four styles together, which is considered advantageous in finding a better solution. We took a deeper look at group and team approaches in Chapter 14.

Finally, James examines some of the barriers to thinking that we might have to contend with:

1. Thinking in absolutes

2. A penchant for immediate gratification

3. Either–or thinking

4. Contentment with the status quo

5. Mindless conformity

6. Pure gut, knee-jerk reactions

Thinking in new ways may require changing one or more of the above, and change of this type can be difficult for many people to embrace.

15.10 THINKING FOR A CHANGE

John Maxwell, a very successful writer and motivational speaker, has been quite active in dealing with leadership issues. In a recent book he provides advice on approaches to thinking that are likely to enhance the life and work situations of people in general. His suggestions are the following 11 points [15.15]:

1. Big-picture thinking

2. Focused thinking

3. Creative thinking

4. Realistic thinking

5. Strategic thinking

6. Possibility thinking

7. Reflective thinking
8. Don't accept popular thinking
9. Shared thinking
10. Unselfish thinking
11. Bottom-line thinking

Maxwell does not attempt to tell his readers what to think, focusing instead on the "hows" for thinking that are listed above. He gives numerous examples that help to reinforce his points, and they are rather convincing. Mastering the foregoing skill areas appears to be very much worth one's consideration.

15.11 SOME GENIUS ATTRIBUTES

Several books have been written about geniuses and what they have contributed to society. One perspective is to look at the attributes of geniuses and then try to infer their approaches and ways of thinking about solving various kinds of problems. One particular book does this in order to "discover your genius" [15.16]. Certainly this is an appropriate way to explore what we might learn from geniuses that we might be able to incorporate into our lives.

The 10 approaches suggested by Gelb in the book cited above are listed below, in association with the revolutionary mind defined by the author:

1. Love of wisdom (Plato)
2. An expanding perspective (Brunelleschi)
3. Optimism, vision, and courage (Columbus)
4. A new world view (Copernicus)
5. Power with balance and effectiveness (Elizabeth I)
6. Emotional intelligence (Shakespeare)
7. Pursuit of happiness (Jefferson)
8. Power of observation and Opening of the Mind (Darwin)
9. Harmonizing spirit, mind, and body (Gandhi)
10. Imagination and combinatory play (Einstein)

Gelb makes many suggestions for how to absorb these positive attributes and try to apply them in one's life. One of the simplest approaches, with lots of advantages, is to keep a notebook. This provides an opportunity to keep these ideas in front of you, in a retrievable form, at all times. We need to do more of these types of simple things. Finally, he also suggests a "genius discovery exercise" at the end of his book that can also be very helpful in approaching this topic.

Several other investigators have found the topic of genius, or its equivalent, of sufficient interest to document their findings. Examples include those listed below.

1. *U.S. News and World Report* came out with a special issue that examined the special genius of Einstein, Freud, and Marx [15.17].

2. James Mannion provided an overview of some of the contributions made by what he called "great thinkers" in a variety of categories, including the military world, human psyche, research and healing, inventors and scientists, pop culture, and others [15.18].

3. Harold Bloom gave us his "mosaic of one hundred exemplary creative minds" in a most exemplary book dealing in large measure with writers [15.19].

4. At the other end of the spectrum, James Gleick decided to focus exclusively on the life and science of Richard Feynman [15.20], a Nobel laureate physicist who himself reached out to laypersons through his several books.

15.12   SYSTEMS THINKING

Perspective 9 for thinking outside the box was presented in Chapter 13. It was called the systems approach and addressed ways of thinking that emphasized the systems aspects of building and managing large-scale complex systems. Despite the fact that such an approach seemed to be rather broad and inclusive, it actually was sharply focused on the elements of the systems approach, considerations of alternatives, building and architecting systems, views of architectures, and the central position occupied by the architectural perspective. One can take a significantly broader view, in which case new aspects of systems thinking come on the scene. These aspects are very briefly cited below.

An overview of systems thinking can be found in Michael C. Jackson's excellent book on that subject, with the subtitle "creative holism for managers" [15.21]. This expansive treatise begins by acknowledging the early contributions of von Bertalanffly [15.22] and Norbert Wiener [15.23] to general systems theory and cybernetics, respectively. Having said that, Jackson discusses a dozen topics that define a variety of systems approaches as well as what he calls *creative holism*. The reader who wishes to explore these matters in greater detail may consult the original and very cogent text by Jackson.

REFERENCES

15.1   de Bono, E., Jr. (1971). *The Use of Lateral Thinking*. New York: Penguin Books.

15.2   de Bono, E., Jr. (1971). *Lateral Thinking for Management*. New York: Penguin Books.

15.3   Savransky, S. (2000). *Engineering of Creativity: TRIZ*. Boca Raton, FL: CRC Press.

15.4   Gelb, M. (1998). *How to Think Like Leonardo da Vinci*. New York: Dell Trade Paperback.

15.5   Ackoff, R. (1978). *The Art of Problem Solving*. New York: Wiley.

15.6   Einstein, A. (1954). *Ideas and Opinions*. New York: Dell Publishing Company.

15.7 Thorpe, S. (2000). *How to Think Like Einstein*. Naperville, IL: Sourcebooks.

15.8 Harvard Business Review (1977). *Harvard Business Review on Breakthrough Thinking*. Boston: Harvard Business Review Paperback.

15.9 Drucker, P. (1977). "The Discipline of Innovation," in *Harvard Business Review on Breakthrough Thinking*. Boston: Harvard Business Review Paperback.

15.10 Nadler, G., and S. Hibino (1990). *Breakthrough Thinking*. Rocklin, CA: Prima Publishing and Communications.

15.11 Csikszentmihalyi, M. (1996). *Creativity*. New York: HarperCollins.

15.12 Finamore, F. (ed.) (1999). *Half Hours with the Best Thinkers*. New York: Gramercy Books, pp. 75–83, from René Descartes (1596–1650).

15.13 Descartes, R. (2004). *Discourse on Method*. New York: Barnes & Noble Books.

15.14 James, J. (1996). *Thinking in the Future Tense*. New York: Touchstone.

15.15 Maxwell, J. (2003). *Thinking for a Change*. New York: Warner Books.

15.16 Gelb, M. (2002). *Discover Your Genius*. New York: HarperCollins.

15.17 *U.S. News & World Report*, Secrets of Genius, Three Minds That Shaped the Twentieth Century, September 2, 2003.

15.18 Mannion, J. (2003). *The Everything Great Thinkers Book*. Avon, MA: Adams Media Corporation.

15.19 Bloom, H. (2002). *Genius: A Mosaic of One Hundred Exemplary Creative Minds*. New York: Warner Books.

15.20 Gleick, J. (1992). *Genius: The Life and Science of Richard Feynman*. New York: Pantheon Books.

15.21 Jackson, M. (2003). *Systems Thinking*. Chichester, West Sussex, England: Wiley.

15.22 von Bertalanffy, L. (1968). *General System Theory*. Harmondsworth, Middlesex, England: Penguin Books.

15.23 Wiener, N. (1948). *Cybernetics*. New York: Wiley.

Chapter **16**

Final Thoughts and a Test

As we continue to move through the twenty-first century, we can expect systems to be more expansive, effective, and complex. Legacy systems will be updated and upgraded, with pressure to create more complex systems through the integration of "stovepipes." These new systems may well be called systems of systems or families of systems or federations of systems, but the trend seems clear. The engineering community will have to cope with these types of changes, in terms of both engineering processes and management. At the same time, software is likely to be more and more important and pervasive, as we have seen, for example, with the increasing use of computers in automobiles. These types of changes may well require new ways of thinking in order to meet the attendant challenges and to achieve the results that we desire in terms of cost-effective systems.

At the top level, this book addresses the matter of building and managing large and complex systems by introducing nine specific perspectives for thinking outside the box. These perspectives have been tried many times before, with considerable success. On this basis, they are recommended by the author as approaches that might be used on a variety of problems that involve systems of the future. One might also think of them as tools for problem solving, moving away from conventional wisdom that has not worked to approaches that might work more effectively.

In Chapters 5 through 13 we have presented and discussed the nine suggested perspectives:

Managing Complex Systems: Thinking Outside the Box, By Howard Eisner
Copyright © 2005 John Wiley & Sons, Inc.

1. **Broaden and generalize**. Expand one's view of a problem and possible solutions.

2. **Crossover**. Move freely from one domain to others to create solutions and leverage.

3. **Question conventional wisdom**. Do not let old and perhaps outmoded approaches get in the way of a fresh look at the problem.

4. **Back of the envelope**. Look for short-form answers that might be adequate by themselves or might steer you toward a more complex solution.

5. **Expanding the dimensions**. Define and look at an issue in all its dimensions.

6. **Obversity**. Explore the obverse side of the matter at hand and the clarity that such an approach offers.

7. **Remove constraints**. Systematically pare away constraints that might be artificial, with the objective of seeing possible solutions that could be cost-effective.

8. **Thinking with pictures**. Use human abilities for observation to sharpen one's view of a problem as well as approaches to its solution.

9. **The systems approach**. Use the body of knowledge in systems theory and practice that increasingly affords us greater capability to build and manage complex systems.

It is suggested that using one or more of the above, in a consistent and continuous manner, will make more likely the activity popularly known as "thinking outside the box."

16.1 MORE EXTRAORDINARY THINKERS

Table 4.1 was a brief list of scientists and inventors that made extraordinary contributions going back hundreds of years. Table 16.1 lists a dozen more contemporary researchers and practitioners who have made significant contributions to thinking about as well as building systems of various types.

TABLE 16.1   Short List of Contemporary Thinkers and Contributors

Contributor	Areas of Contribution
Russell Ackoff	Management and operations research
Norman Augustine	Engineering executive; building systems
Barry Boehm	Software economics and research
Peter Drucker	How business works and should work
Richard Feynman	Physics Nobel laureate and extraordinary teacher
Jay Forrester	System dynamics
Michael C. Jackson	Systems thinking
Alan Pritsker	Modeling and simulation software
Eberhardt Rechtin	Architecting of systems and organizations
Andrew Sage	Theory/practice of systems engineering
Herbert Simon	Artificial intelligence; Nobel laureate
Norbert Wiener	Cybernetics

We continue to try to understand how all these contributors did what they did, but that is not a straightforward matter. As suggested in various parts of this book, there are many ways of thinking about knotty problems—many approaches to consider. In the final analysis we have to pick and choose what works for us as individuals and what might work in a group setting. Fortunately, most of our best thinkers have given us some idea, through their writings, of how they looked at and solved the problems they tackled and wrestled their way through the unknowns.

16.2 IS THERE LIFE OUTSIDE THE BOX?

In an earlier chapter we alluded to the possibility that not everyone is ready to accept thinking that may be outside the box. If such thinking is indeed largely the province of only 5% of people, 19 out of 20 people are thinking in other directions. They often will not look kindly on the "better" idea and may be inclined to reject it without due consideration. This can be frustrating for the unconventional thinker, who in today's world needs to adjust to that type of reality. The recommended approach here is to be patient and try to maintain a steady and persistent attitude and response. And if you're getting lots of resistance from just about everyone, another examination of your own ideas may well be in order. If you're in an environment that rejects most of what you have to offer as an independent thinker, it may well be time to move on.

Many of the lessons learned in this regard can be found in the literature having to do with creativity and innovation. There have been many reports of what it took finally to be heard in a corporate setting, or situations in which top management just plain missed the boat. An example of the latter, often cited, is the case of Xerox PARC. There are numerous examples of that type. At the same time, also with reference to Xerox, there is the wonderful story of how they bet everything on an idea and a technology that they came to believe in and fully embrace. Apparently, thinking that xerography would work and find a massive long-term market was a long way outside the box. These types of success stories keep us engaged even though there may be many hurdles to jump over.

16.3 TEST YOURSELF

I have prepared a questionnaire that should give readers some insights into possible tendencies to move outside the box and be successful in that endeavor. Readers are invited to take the test now, and look at the results afterward.

TDP Questionnaire

Select the one statement from each pair of statements that appears to be more appropriate.

1. a. Most of the time I am content to go with the flow.
   b. I am able to work my way past obstacles.
2. a. I tend to be an original thinker.
   b. I value consensus and agreement on most issues.
3. a. I value consensus and agreement on most issues.
   b. I take the initiative most of the time.
4. a. I like to get to the essence of a subject.
   b. I am diligent in pursuit of a goal.
5. a. Most of the time I am content to go with the flow.
   b. I tend to be an original thinker.
6. a. I am able to work my way past obstacles.
   b. I often question conventional wisdom.
7. a. I take the initiative most of the time.
   b. I do not like to stand out in a group.
8. a. I like to get to the essence of a subject.
   b. Most of the time I am content to go with the flow.
9. a. I like to get to the essence of a subject.
   b. I am able to work my way past obstacles.
10. a. I often question conventional wisdom.
    b. I take the initiative most of the time.
11. a. I value consensus and agreement on most issues.
    b. I like to get to the essence of a subject.
12. a. I am diligent in pursuit of a goal.
    b. I do not like to stand out in a group.
13. a. I do not like to stand out in a group.
    b. I am able to work my way past obstacles.
14. a. I often question conventional wisdom.
    b. I value consensus and agreement on most issues.
15. a. I am diligent in pursuit of a goal.
    b. I value consensus and agreement on most issues.
16. a. I take the initiative most of the time.
    b. I tend to be an original thinker.
17. a. Most of the time I am content to go with the flow.
    b. I am diligent in pursuit of a goal.
18. a. I like to get to the essence of a subject.
    b. I do not like to stand out in a group.
19. a. I am diligent in pursuit of a goal.
    b. I tend to be an original thinker.

20. a. I often question conventional wisdom.
    b. Most of the time I am content to go with the flow.
21. a. I am able to work my way past obstacles.
    b. I often question conventional wisdom.
22. a. I take the initiative most of the time.
    b. I like to get to the essence of a subject.
23. a. I do not like to stand out in a group.
    b. I tend to be an original thinker.
24. a. I am diligent in pursuit of a goal.
    b. I value consensus and agreement on most issues.
25. a. I am able to work my way past obstacles.
    b. I tend to be an original thinker.
26. a. Most of the time I am content to go with the flow.
    b. I take the initiative most of the time.
27. a. I do not like to stand out in a group.
    b. I often question conventional wisdom.

The scorecard for the questionnaire is provided in the following section. Go to it now and obtain your score on the questionnaire.

16.3.1 Discussion of Scores for Questionnaire

The questionnaire, using your answers, develops a profile of your tendencies along three dimensions:

1. Moving outside the box in your thinking (TOTB)
2. Discomfort with sticking your neck out (DISC)
3. Persistence in pursuit of your goal (PERS)

Your scores in each of these three areas can be obtained from the scorecard that follows.

From the way the questionnaire is structured, the maximum score that can be obtained in any one of the three dimensions is 18. A completely "balanced" score would be 9–9–9. Considerably lower than a 9 (such as a 6) in any dimension would be considered a tendency to avoid that area. On the other hand, a scale of 12 or more in any area indicates a strong affinity for that mode of behavior or point of view. High scores under TOTB and PERS suggest that you already think outside the box, and are quite persistent, respectively. A high score under DISC indicates that you don't like to rock the boat, so that even if you think outside the box, you may not do the necessary follow-up to support your ideas. Beyond that, you can make all the necessary interpretations. Hopefully, this will be helpful in assessing your own tendencies with respect to these particular dimensions. What should you do about them? Only you are able to answer that question.

Scorecard for Questionnaire

Transcribe questionnaire answers to this scorecard by circling appropriate selections. *Example*: If you selected 1a on the questionnaire, circle 1a here under PERS. If you selected 1b on the questionnaire, circle 1b here under TOTB. Then add up the number of circles, under each column, and place the answer for each column following the word "Sum."

TOTB	DISC	PERS
2a	1a	1b
4a	2b	3b
5b	3a	4b
6b	5a	6a
8a	7b	7a
9a	8b	9b
10a	11a	10b
11b	12b	12a
14a	13a	13b
16b	14b	15a
18a	15b	16a
19b	17a	17b
20a	18b	19a
21b	20b	21a
22b	23a	22a
23b	24b	24a
25b	26a	25a
27b	27a	26b

Sum

Score for Thinking	Score for Discomfort	Score for Persistence

(Verify that total of sums add to 27)

16.4 FINAL WORDS

If applied diligently, the nine thinking perspectives suggested in this book are likely to introduce you to new pathways to success in coping with difficult problems in relation to building and managing complex systems. Modifying one's approach to problem solving has been known to be challenging. However, it is the type of change that can bring very satisfying rewards. Readers are urged to be patient in trying these perspectives, and when necessary, to go back to the various discussions and examples for reinforcement. Try to remember the story about how one can find one's way to Carnegie Hall: practice, practice, practice.

Index

Managing Complex Systems: Thinking Outside the Box, By Howard Eisner
Copyright © 2005 John Wiley & Sons, Inc.